BIRDS *of* STONE

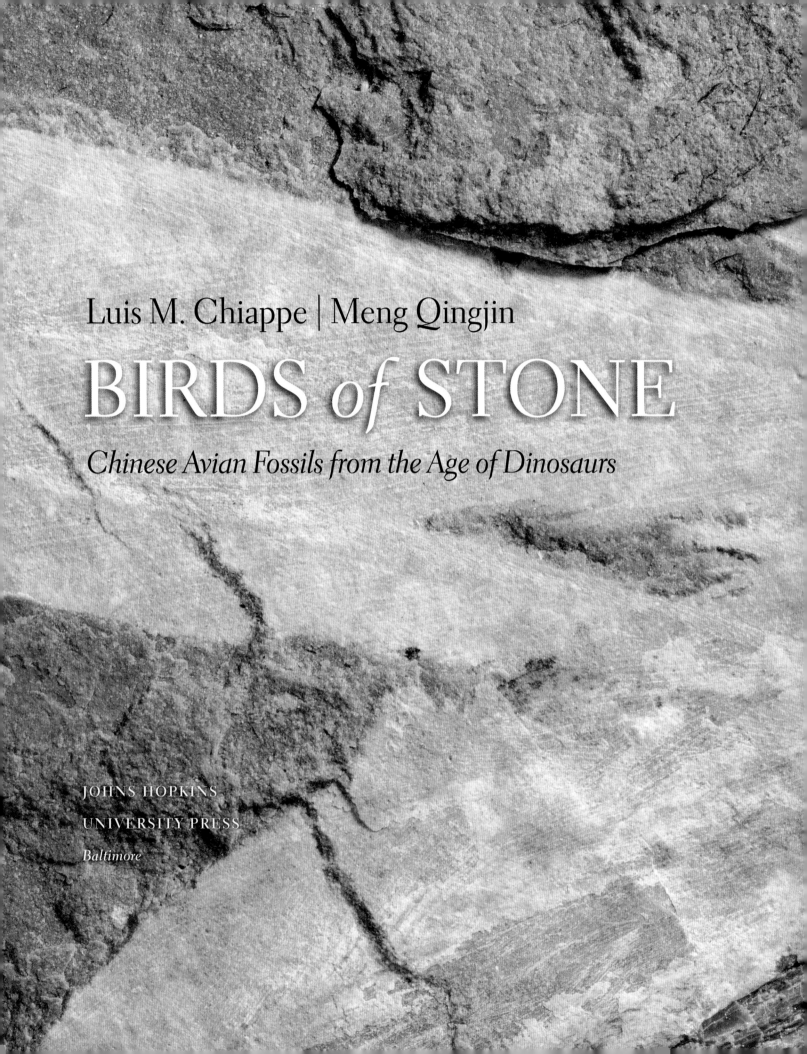

Luis M. Chiappe | Meng Qingjin

BIRDS *of* STONE

Chinese Avian Fossils from the Age of Dinosaurs

JOHNS HOPKINS
UNIVERSITY PRESS
Baltimore

© 2016 Johns Hopkins University Press
All rights reserved. Published 2016
Printed in China on acid-free paper
9 8 7 6 5 4 3 2 1

Johns Hopkins University Press
2715 North Charles Street
Baltimore, Maryland 21218-4363
www.press.jhu.edu

Library of Congress Cataloging-in-Publication Data

Names: Chiappe, Luis M. | Meng, Qingjin, 1962–
Title: Birds of stone : Chinese avian fossils from the age of dinosaurs /
    Luis M. Chiappe and Meng Qingjin.
Other titles: Chinese avian fossils from the age of dinosaurs
Description: Baltimore : Johns Hopkins University Press, 2016. | Includes
    bibliographical references and index.
Identifiers: LCCN 2015037774| ISBN 9781421420240 (hardcover : alk. paper) |
    ISBN 9781421420257 (electronic) | ISBN 1421420244 (hardcover : alk. paper) |
    ISBN 1421420252 (electronic)
Subjects: LCSH: Birds, Fossil. | Paleontology—China.
Classification: LCC QE871 .C44 2016 | DDC 568.0951—dc23 LC record
    available at http://lccn.loc.gov/2015037774

A catalog record for this book is available from the British Library.

Special discounts are available for bulk purchases of this book. For more information,
please contact Special Sales at 410-516-6936 or specialsales@press.jhu.edu.

Johns Hopkins University Press uses environmentally friendly book materials,
including recycled text paper that is composed of at least 30 percent post-consumer
waste, whenever possible.

To our sons,

Luca and Fanhao,

who gave us wings

to see life from the sky.

ACKNOWLEDGMENTS

There are many people who need to be acknowledged; however, this book would not have been possible without the hard work of Maureen Walsh and Stephanie Abramowicz, who spent countless weeks in China, preparing and photographing specimens, respectively—to both Maureen and Stephanie we owe the beauty of this book. Many colleagues and museum officials provided access to the spectacular fossils featured in this book. These individuals include Liu Linde, Liu Di, Zeng Zhaohui, and Wang Yu of the Beijing Natural History Museum; Ji Qiang, Wang Xuri, Ji Shu'an, and Lu Junchang of the Chinese Academy of Geological Sciences; Zhou Zhonghe, Zhang Fucheng, Jingmai O'Connor, and Wang Min of the Institute of Vertebrate Paleontology and Paleoanthropology; Gao Chunling, Zhang Fengjiao, and Zhao Bo of the Dalian Natural History Museum; Zhang Hua and Chang Huali of the Henan Geological Museum; Sun Ge and Hu Dongyu of the Liaoning Provincial Museum (Shenyang Normal University); Sun Dewei, Xu Xiaoyin, Zhang Nanjun, and Wu Qin of the Changzhou Dinosaurland Co., Ltd.; Zheng Xiaomei and Zheng Xiaoting of the Shandong Tianyu Museum of Natural History; Teng Fang Fang of the Dalian Museum of Prehistory; and Zhang Zihui of the Beijing Normal University. To all of them, we extend our deepest gratitude. Several people also played key roles as facilitators, hosting the US team in China and providing logistical support; in this regard we are very grateful to Ji Qiang, Wang Xuri, and Liu Linde.

Several employees of the Natural History Museum of Los Angeles County and conservation consultants need to be acknowledged for additional preparation, molding, and casting of specimens: Doug Goodreau, Doyle Trankina, Aisling Farrell, Gary Takeuchi, and Marilyn Fox. Additionally, many colleagues enriched this book through academic discussions and research collaborations. Particularly, we want to thank Jesús Marugán-Lobón, Alyssa Bell, Wang Min, Wang Xuri, Anusuya Chinsamy, Lars Schmitz, Francisco Jose Serrano Alarcón, Guillermo Navalón, Michael Habib, Justin Hall, and Jingmai O'Connor.

Helmut Tischlinger, Jeremy Cook, Mick Ellison, Anusuya Chinsamy, Wang Min, Liu Xinzheng, Timothy Rowe, Amanda Falk (with research supported by grant number NRICH-1105-A21F from the National Research Institute of Cultural Heritage [Korean Geodiversity and Fossil Site Research]), Xu Xing, Jingmai O'Connor, Liliana D'Alba, and Lorraine Meeker provided a number of the images featured in this book—we thank them all very much. We are also grateful to Tyler Hayden, Maureen Walsh, and Lisa Granados, who provided instrumental support editing and organizing the manuscript.

Our travels and research were supported by many sources, including multiple grants from the National Natural Science Foundation of China, the Institute of Geology at the Chinese Academy of Geological Sciences, the Ministry of Land and Resources of The Peoples Republic of China, the Beijing Academy of Science and Technology, the Frank Chapman Memorial Fund of the American Museum of Natural History, the Eppley Foundation, the Quest Program of the Discovery Channel, and the National Science

Foundation. Doreen and Glenn Gee, Carl and Lynn Cooper, Ron and Judy Perlstein, and Richard and Eileen Garson generously provided additional support. Finally, we are in debt to Vincent Burke from Johns Hopkins University Press, who supported this project all along, and to Juliana McCarthy, Maria denBoer, Mary Lou Kenney, and Tracy Baldwin for their editorial and design oversight.

The illustration of the chicken embryo on page 114 is from a photograph by Dr. Jeremy Cook from a specimen prepared in the laboratory of Professor Ruth Bellairs. It appeared on the cover of *The Atlas of Chick Development*, by Ruth Bellairs and Mark Osmond (Academic Press, 1998), and is reproduced with permission.

CONTRIBUTING INSTITUTIONS

BMNHC	Beijing Museum of Natural History, Beijing, China
CDL	Changzhou Dinosaur Park, Changzhou, China
CNUVB	Beijing Normal University, Beijing, China
DNHM	Dalian Natural History Museum, Dalian, China
GMV	Geological Museum of China, Beijing, China
HGM	Henan Geological Museum, Zhengzhou, China
IVPP	Institute of Vertebrate Paleontology and Paleoanthropology, Beijing, China
LACM	Natural History Museum of Los Angeles County, Los Angeles, California, USA
LPM	The Museum of Beipiao, Sihetun, Chaoyang, Liaoning Province, China
NIGPAS	Nanjing Institute of Geology and Paleoanthropology, Nanjing, China
PKUP	Peking University, Beijing, China
SDM	Shandong Museum, Jinan, China
STM	Shandong Tianyu Museum of Nature, Pingyi, China
XHPM	Xinghai Museum of Prehistoric Life of Dalian, Dalian, China

BIRDS of STONE

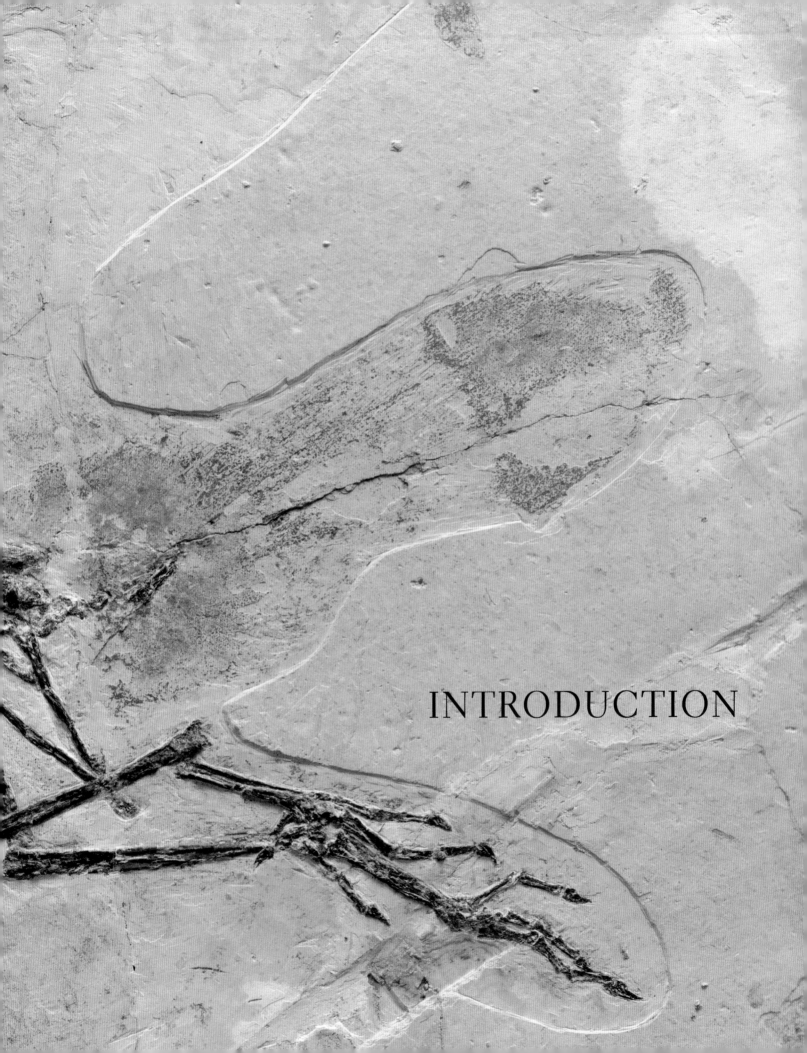

INTRODUCTION

It was very cold. The countryside was blanketed in snow, intermittently dotted by drab houses and small villages. We were traveling in a van to Chaoyang, an ancient city and birthplace of Chinese culture, some five hours northeast of Beijing, in the western portion of Liaoning Province. What brought us there was the promise of a treasure trove of birds of stone: a large collection of ancient fossils housed in a small local museum. Since the early 1990s, a wealth of extraordinary fossils of birds and other organisms had been unearthed from this region in the northeastern corner of China. Dating back more than 120 million years, these spectacular fossils were rewriting the evolutionary history of many groups of animals and plants in deep time.

Chaoyang gave us a glimpse of what the city might have been in its heyday, centuries ago. Tall brick pagodas with Buddhist carvings stood out from a maze of ornate classic buildings and market alleyways. The museum was located in one of these buildings, its many fossils displayed inside delicate glass cases or hung in picture frames decorating the walls. We had arrived to find what we were after, a remarkable collection of archaic birds, most of them still cloaked with feathers and preserving delicate details of their skeletons. What we could not have predicted was that this magnificent collection of ancient birds was only the beginning. Since that time, thousands of exquisite fossils like these have been dug up from a vast expanse in northeastern China—in western Liaoning Province, north-eastern Hebei Province, and southeastern Inner Mongolia—and they have provided the clearest window into the earliest stages of the evolution of birds.

Trapped in stone, the delicate remains of thousands of ancient birds that lived more than 120 million years ago have been unearthed from numerous sites across a vast territory in northeastern China. These amazing fossils are rewriting our understanding of the early evolutionary history of birds.

Birds have a rich and ancient past; their origins date back to the Jurassic Period, more than 150 million years ago, deep within the Age of Dinosaurs, or the geologic time interval called the Mesozoic Era (~ 252 to 66 million years ago). Over millions of years, these animals have evolved a stunning array of appearances, behaviors, and adaptations: a 2-gram (0.07 ounce) hummingbird beats its wings 80 times a second, migrating geese fly above Mount Everest at 9,000 meters (nearly 30,000 feet), gannets gracefully strike the surface of the ocean after a 40-meter (130 feet) plunge, and the extravagant elegance of a fully plumed bird-of-paradise is unmatched. During such a long evolutionary history, birds have thrived on every continent and occupied most habitats, including the frigid poles and the fiercest deserts.

With more than 10,000 living species, birds are the most species-rich group of land-dwelling vertebrates (animals with a backbone); their number is almost double that of today's mammals. Such a magnificent diversity of shapes, colors, and lifestyles evolved over many millions of years from the ancestors of present-day birds that lived in the utterly different world of the Mesozoic and that ultimately gave rise to their living descendants. The diversity of birds that lived during the Mesozoic was not just limited to the direct forebears of today's birds. The fossil record of this time also documents the origin of an array of primitive lineages of birds that evolved for millions of years alongside their large dinosaurian relatives and that became extinct with them during this geologic era, vanishing without living descendants.

The time involved in the vastly long evolutionary history of birds—from the Jurassic to the present—is sometimes difficult to comprehend. It is the immensity of time over which landscapes have been shaped or which has led to the physical nature of our modern world: the known history of birds is in fact more than 20 times longer than the entire history of the human lineage. The oldest known bird is the famous *Archaeopteryx lithographica*—a feathered, crow-sized animal with sharp claws at the tip of its fingers, a long bony tail, and a mouth full of small, pointy teeth—which lived about 150 million years ago in what is today southern Germany. *Archaeopteryx* lived more than 80 million years before *Tyrannosaurus rex* walked the surface of the Earth. With the discovery of *Archaeopteryx's* fossils in the middle of the nineteenth century, the vastness of the evolutionary saga of birds became manifest. Some other groups of ancient Mesozoic birds were discovered in the subsequent years of that century, including the 1.5-meter-long (5 feet) diver *Hesperornis regalis* and the tern-sized, flighted *Ichthyornis dispar*, the fossilized remains of which are abundant in chalk formed at the bottom of a shallow sea that 85 million years ago covered today's western Kansas and other portions of central North America. These early discoveries revealed the antiquity of the bird lineage and hinted at the diversity of birds that shared the world with the fearsome dinosaurs of the Mesozoic Era.

Despite the importance of these discoveries and of others during most of the twentieth century, the true diversity of the birds that lived during the Mesozoic remained elusive and the available evidence for understanding many aspects of their evolutionary history scant. Toward the end of the millennium, such was the dearth of the avian fossil record of the Mesozoic Era that most scientists believed that, in the shadow of the large dinosaurs, bird diversity had remained marginal and that it was not until the extinction of these behemoths that the group flourished.

Discoveries worldwide since the 1990s have helped elucidate the rich history of birds in an unprecedented way. Numerous fossils of Mesozoic birds have been unearthed from sites around the world, but nowhere in such abundance, diversity, or superb preservation

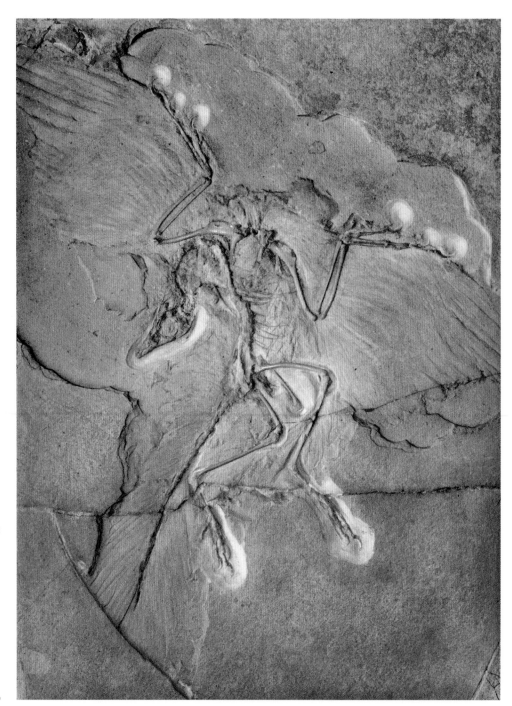

An icon of evolution, the Berlin specimen of *Archaeopteryx lithographica* is perhaps the most celebrated of all fossils. Dating back to about 150 million years ago, this and other *Archaeopteryx* specimens constitute the earliest known evidence of birds.

as in northeastern China. Thousands of exquisite fossils have been collected during the past three decades there. This remarkable fossil aviary, together with the well-preserved remains of many other animals and plants, is known as the Jehol Biota, a historic reference to the name of the ancient region that centuries ago was the seat of the powerful Khitan Empire. Dating between 131 and 120 million years ago, a time interval contained within the first half of the Cretaceous Period of the Mesozoic Era, the Jehol Biota provides a unique window into a terrestrial ecosystem at a time when animals and plants critical for understanding modern life show increased evolutionary diversification and when the geography of the world was vastly different from that of today.

The beige, ochre, and gray fine-grained shales that contain the Jehol fossils record the structures of these organisms in astonishing detail. Soft tissues such as hair, feathers, and skin often surround the skeletons of many animals and even the most delicate structures such as the wings of insects and the veins of leaves are exquisitely preserved. These extraordinary fossils tell us of a thriving ecosystem of tall forests rimming a system of interconnected lakes that developed under a temperate climate of marked seasonality. This idyllic environment was occasionally impacted by devastating eruptions of neighboring volcanoes that led to events of ecological collapse and mass mortality, with the remains of the many animals and plants that lived in this thriving ecosystem sinking to the bottom of the lakes, where they were gently covered by the settling sediments.

The amazing ancient menagerie captured by these drastic events has left us with a wealth of fossils that are rewriting the evolutionary history of many groups of organisms, from birds to mammals and from insects to flowering plants. Such paleontological treasures have revealed that early in their history, birds evolved a stunning diversity of shapes and acquired many different adaptations. But the Jehol Biota has done much more than document an unexpected diversity of ancient birds filling a vast gap in the fossil record that separates living birds from their extinct dinosaurian predecessors. These spectacular fossils have radically transformed our understanding of the lives of birds that span a large portion of the avian family tree, from near its base to the branches close to the origin of modern birds. Indeed, these discoveries have detailed one of the most astounding evolutionary transitions in the history of life and revealed a more thorough image of how birds became the fascinating animals they are.

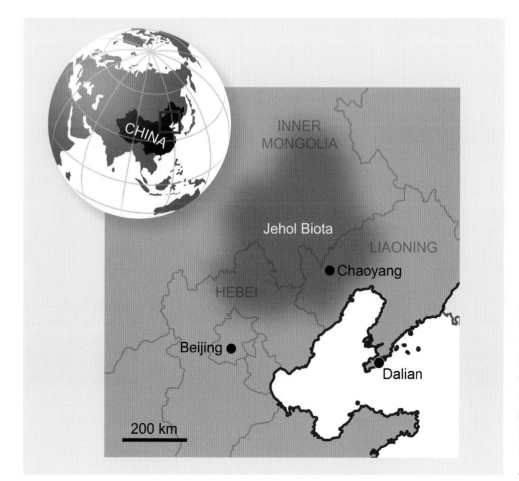

Dozens of quarries containing fossils of the renowned Jehol Biota poke into the land over a large region that includes western Liaoning Province, northeastern Hebei Province, and the eastern portion of the Inner Mongolia Autonomous Region in northeastern China.

8

The Jehol Biota offers an exceptional window into ancient life that existed during the Cretaceous Period, following the first appearance of birds in the fossil record. The remarkable Jehol aviary is contained within three geological formations: the Huajiying Formation and its equivalents (approximately 131 million years ago), the Yixian Formation (from about 129 to 122 million years ago), and the Jiufotang Formation (roughly 122 to 120 million years ago).

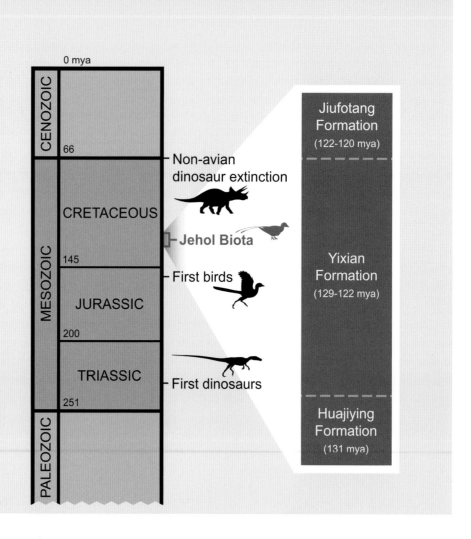

Not long ago, we thought that during their deep evolutionary history, birds had only a fraction of the diversity that they boast of today. China's newly discovered ancient menagerie has transformed our understanding of the kinds of birds that lived during the Mesozoic. The many discoveries from the Jehol Biota now join a wealth of fossil finds worldwide indicating that avian diversity during the Age of Dinosaurs might have been as colorful and extraordinary as it is today. While the exquisite fossils of the Jehol Biota still present us with an incomplete roster of what this early avian diversity might have been, they delight us with their beauty and insight. This book is an attempt to deliver a pictorial narrative of such amazing diversity and to highlight how these remarkable fossils bring our understanding of the early evolution of birds to new heights.

Virtually intact and still surrounded by its plumage, this 125-million-year-old fossil of *Confuciusornis sanctus* is one of thousands of exquisite fossil birds whose corpses became sealed by settling sediments at the bottom of an ancient lake in the Jehol region of northeastern China.

THE AVIAN FOSSILS *of* JEHOL

A wealth of exquisitely preserved fossils has been unearthed from multicolored shales at various quarries around Sihetun (*left*), a village in the countryside of Liaoning Province, some 400 kilometers (250 miles) northeast of Beijing. This remarkable treasure trove includes a variety of feathered non-avian dinosaurs and early birds, and a myriad of other life forms that lived with them. Fossils from these quarries, together with thousands of others from similar sites across a wide region of northeastern China, make up the celebrated Jehol Biota. These spectacular fossils, and the rocks that contain them, provide the world's clearest window into terrestrial ecosystems that thrived during the first half of the Cretaceous Period between 131 and 120 million years ago.

   Buried at the bottom of lakes that used to dot a forested landscape, vestiges of the animals and plants that inhabited these ancient ecosystems became compressed by the weight of mud, silt, and ash that settled on the lake floor over millennia. Their fossilized remains developed a plane of weakness, an irregularity in the otherwise continuous column of lake sediments, which enabled the rock slabs to be split in half, often revealing two nearly identical mirror images reflected in stone (*below*). In recent decades, intensified quarrying of these fine-grained rocks (*overleaf*) has produced an amazing array of fossils, and their discoveries are transforming our understanding of the evolution of many groups of organisms.

Eocathayornis walkeri
IVPP-V10916A\B
Jiufotang Formation
Boluochi, Chaoyang County
Liaoning Province

The fine sediments preserving the exquisite fossil birds of the Jehol Biota are contained in three separate geologic formations, or discrete packages of rocks, which are defined by specific characteristics. Like the pages of a thick book, the hundreds of meters of shale and other rock layers that make up these geologic formations recount the history of the lakes and their surrounding forests that prevailed in the ancient Jehol landscape. Radioactive elements locked inside these rocks tell us that sediments of the oldest, the Huajiying Formation, settled about 131 million years ago. The intermediate Yixian Formation has been dated to about 129 to 122 million years ago, and the rocks of the Jiufotang Formation, the youngest geologic unit containing Jehol fossils, are thought to be 122 to 120 million years old.

Among the earliest Jehol fossils, including the small enantiornithine shown here (*right*), are the oldest known Cretaceous birds worldwide. Yet, these ancient birds are nearly 20 million years younger than the famous *Archaeopteryx lithographica*—the world's oldest bird species—whose carcasses became buried in the coastal lagoons that toward the end of the Jurassic rimmed what is now southern Germany. This enormous temporal gap in the fossil record of early birds, the millions of years separating *Archaeopteryx* from the earliest Jehol birds, represents the most challenging frontier in avian paleontology. Fossils from this dark period of the early history of birds are sure to contain critical evidence documenting the initial phases of these animals' evolution.

In the absence of a fossil record of the birds that lived during the first million years following the time of *Archaeopteryx*, the paleontological bonanza from the Jehol region of northeastern China has become paramount for understanding all aspects of the early evolution of birds. These amazing fossils speak of an enormous diversity of birds thriving early in the Cretaceous and they tell us of many evolutionary innovations arising at the dawn of the group. The Jehol birds thus lived at a critical juncture in the ancient evolutionary saga of birds, and the information they provide for clarifying previously obscure chapters of the evolution of these animals remains unparalleled.

Enantiornithine indeterminate
BMNHC-PH1154B
Huajiying Formation
Sichakou, Fengning County
Hebei Province

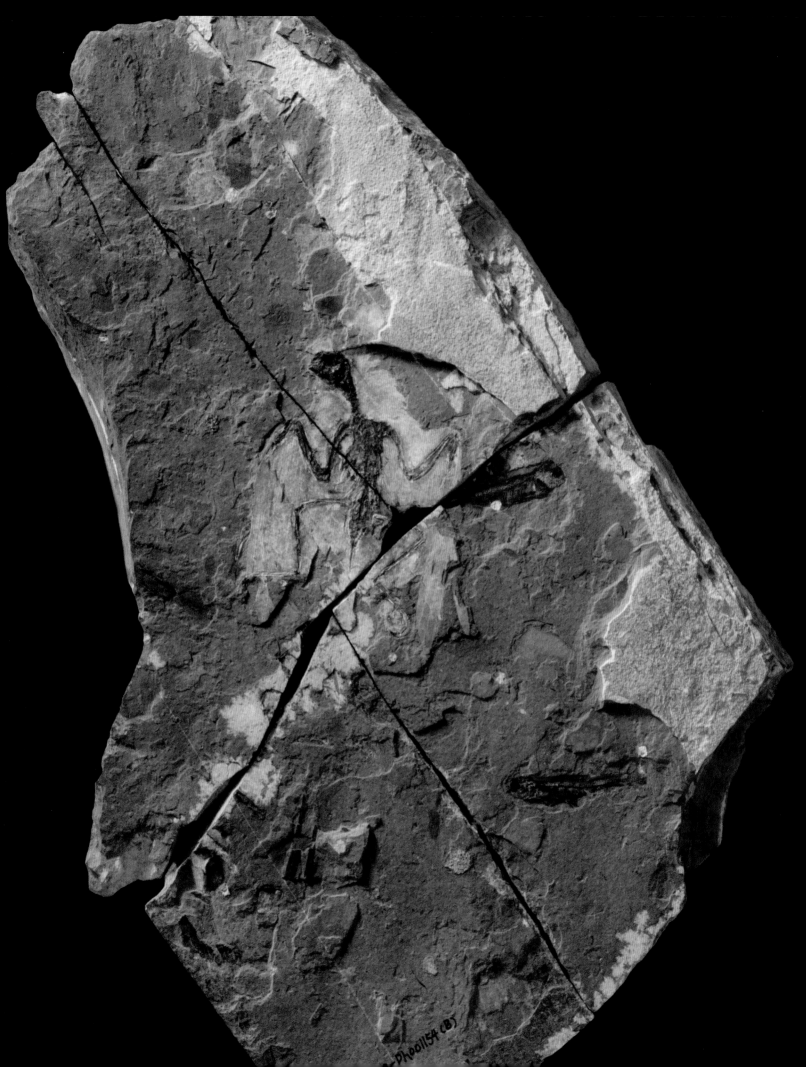

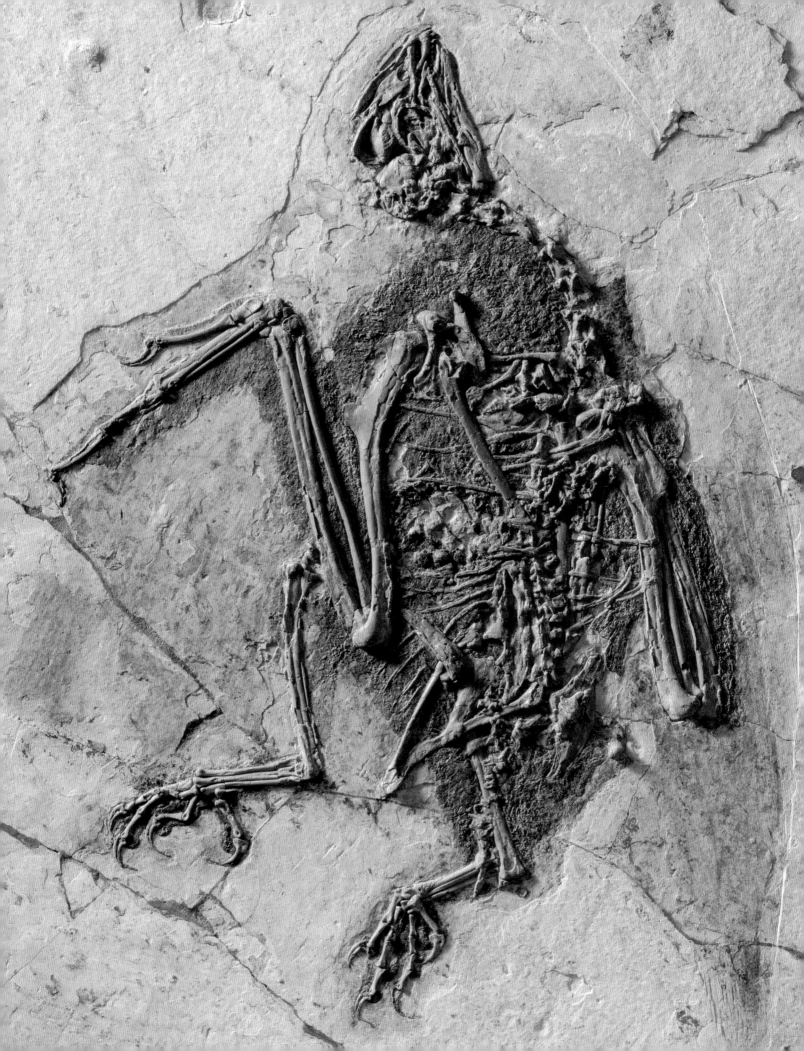

Under rare conditions, the fine sediments that buried the carcasses of the ancient Jehol birds enabled even their softest parts to be preserved in extraordinary detail. Most of these fossils are virtually complete, their corpses undisturbed, with their bones retaining their natural joints and their skeletons surrounded by large portions of plumage. In some instances, the outlines of the skin and flesh are still clearly visible in the neighboring rock. A blackened mass surrounding the skeleton of one fossil (*left*) replicates the wavy folds of skin that formed the fleshy wing, the contours of the shank and thigh, and the large pads that cushioned the joints of the toes; remnants of the wings' primary and secondary flight feathers are also visible. Minute aspects of the structure of feathers are also regularly preserved in many other Jehol birds and so are the horny sheaths that encased the bony claws, the hard scales covering the foot and toe bones, and the reticulate skin that separated the toes (*overleaf*).

In the presence of countless of these amazing fossils, it is easy to forget that these animals lived many millions of years ago. Undoubtedly, an infrequent combination of special conditions must have acted together for such exceptional preservation to take place. Rapid entombment, minimal water circulation, and the virtual absence of oxygen at the lake floor prevented the destructive action of scavengers and hindered the natural process of microbial decay. More than 100 million years ago, as the carcasses of many of the birds that lived in and around the Jehol forests settled on the bottom of ancient lakes, their virtually intact corpses were rapidly sealed and compressed by the weight of sediments. The prevailing conditions at the bottom of the lake allowed their bodies to be pristinely preserved, which as fossils have enlightened our understanding of some of the earliest phases of the evolutionary history of birds.

Sapeornis chaoyangensis
HGM-41HIII0405
Jiufotang Formation
Dapingfang, Chaoyang County
Liaoning Province

Yanornis martini
STM-9-5
Yixian or Jiufotang Formation
Jianchang County
Liaoning Province

Sapeornis chaoyangensis
DNHM-D3078
Yixian Formation
Jianchang County
Liaoning Province

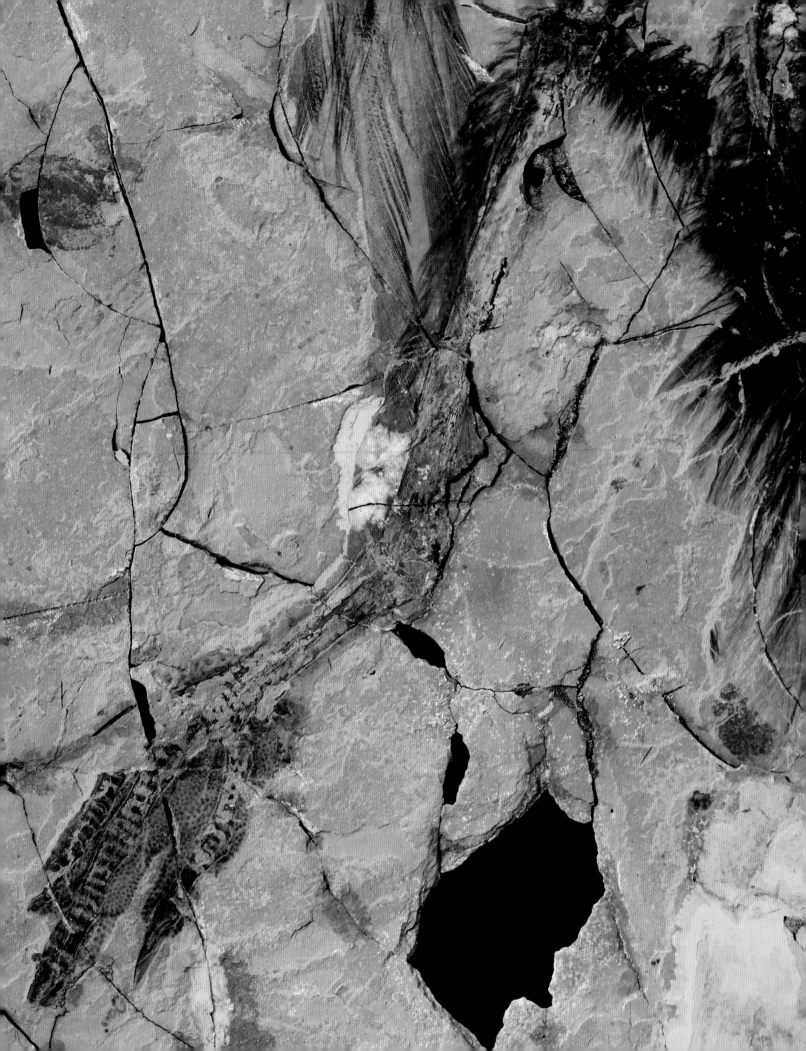

The exceptional preservation of soft tissues in the Jehol fossil birds extends beyond what we can see with the naked eye. In many cases, ultraviolet (UV) illumination allows us to visualize features that are difficult to see under normal light (*overleaf*). The delicate skin and muscles that made up the fleshy wings and legs are often revealed under such glowing light (*left* and *below*). The shape of these tissues indicates that the same main skinfolds and muscles that characterize the wings and legs of living birds were already present in their bygone relatives. They also show that the feet of these ancient birds were devoid of main muscles, as is the case in today's birds as well. Likewise, UV illumination reveals that a narrow portion of flesh encased the short bony tail of some of these animals (*left*), a major difference from the fatty swollen structure that characterizes the end of the tail of living birds (often called the pope's nose). Ultraviolet light has also proven to be a powerful tool for identifying the common "enhancements" the Jehol fossil birds suffer in the hands of people trying to increase their commercial value. An unfortunate number of fossils from these sites find their way into the illegal international market every year. The skeletons often include fabricated additions, painted plumage, and other artificial "improvements," all of which are easily identifiable under the bluish glow of UV light.

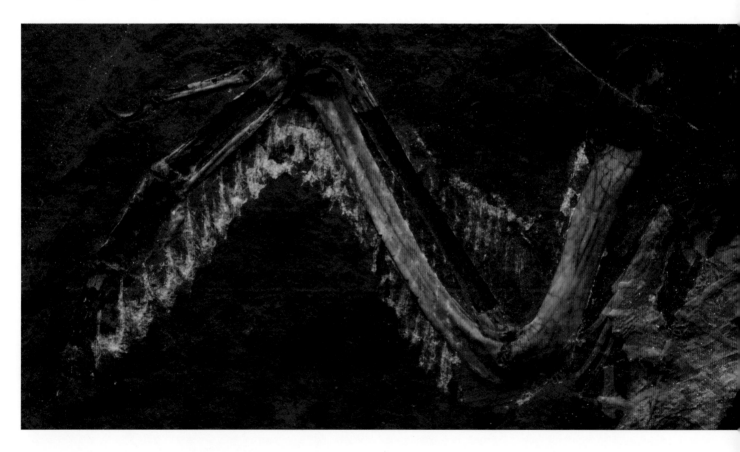

Enantiornithine indeterminate
BMNHC-PH1154A
Huajiying Formation
Sichakou, Fengning County
Hebei Province

Hongshanornis longicresta
DNHM-D2946
Yixian Formation
Lingyuan County
Liaoning Province

Enantiornithine indeterminate
BMNHC-PH1156A
Huajiying Formation
Sichakou, Fengning County
Hebei Province

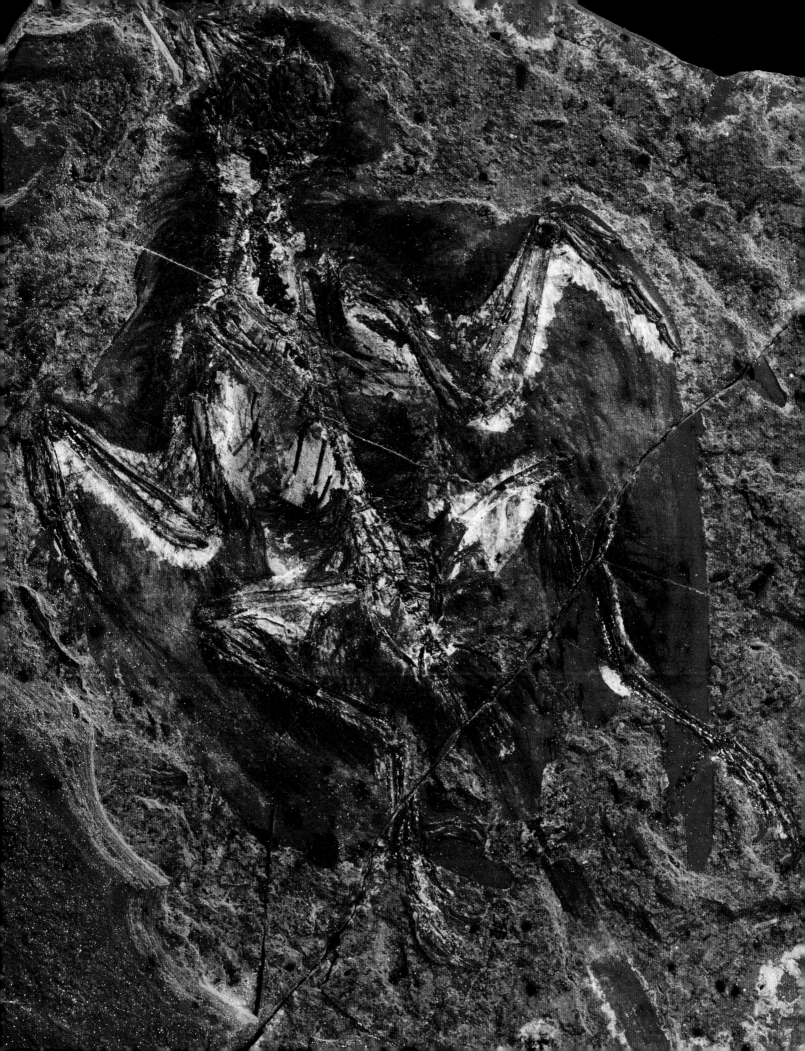

Well-documented interactions involving vertebrates and insects are extremely rare in the fossil record. The Jehol Biota includes an impressive number of different types of insects that lived alongside birds and many other vertebrates. However, only a handful of Jehol fossils provide evidence of the association between the corpses of birds and the insects that fed on their carcasses. One fossil (*right*) reveals the existence of tiny structures—spindle-shaped and often in clusters (*below*)—that resemble the eggs of many living insects. These 120-million-year-old structures are preserved inside the mouth as well as between and near the feet (*box*) of this bird (*right*). Because insect eggs are notoriously similar across different groups, it is difficult to identify the type of insects that laid these ancient eggs. Likewise, it is difficult to say whether they were insects that lived on land or in the water. This notwithstanding, these eggs were probably laid by insects that fed on carrion (flies, beetles, and others) before the corpses sank to the lake's bottom. The information derived from this exceptional fossil is limited but nonetheless important. Such a rare association highlights once again the extraordinary and multileveled preservation of the Jehol Biota, and it promises to provide forensic evidence that may help us elucidate more precisely the settings in which these animals died and were finally entombed.

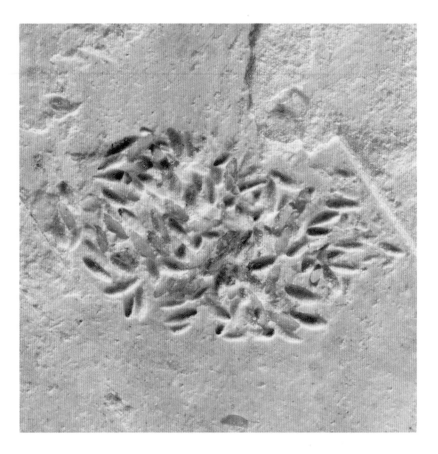

Longipteryx chaoyangensis
BMNHC-PH826
Jiufotang Formation
Dapingfang, Chaoyang County
Liaoning Province

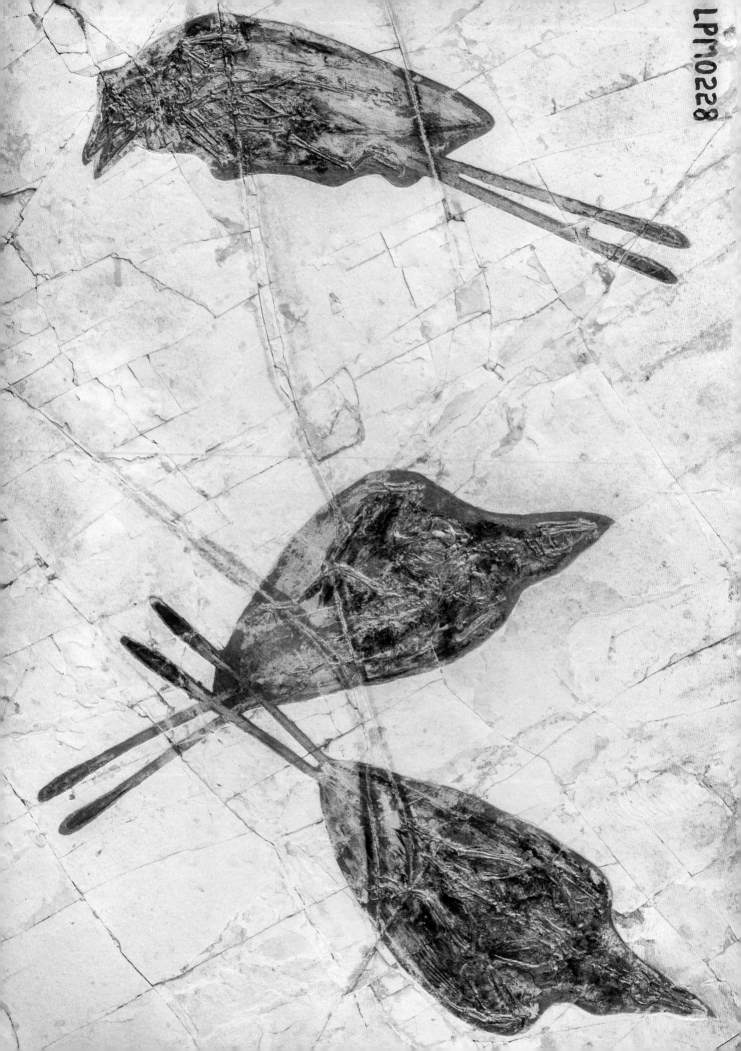

Abrupt changes in the environmental conditions of the time are often interpreted as the cause of the paleontological bonanza of the Jehol Biota. Large accumulations of fossils, sometimes of a single species, point to episodes of mass mortality. Fossils shown here (*left* and *overleaf*) are remarkable examples of how these dramatic events severely impacted the Jehol aviary. Layers of ash and lava, and the nature of the rocks entombing the fossils, tell us that this menagerie lived in the vicinity of active volcanoes. Fierce eruptions releasing a variety of toxic gases are regarded as the smoking gun for such carnage. Details of the sediments and the fossils also point to pyroclastic flows as perpetrators. These deadly clouds of incandescent gas, ash, and other volcanic debris are presumed to have regularly passed through the ancient forests, killing almost everything in their path and carrying countless corpses into the neighboring lakes. Rapid changes in salinity and overall composition of the lake water are envisioned as responsible for the episodic mass mortality among fish (*below*) and other aquatic animals.

Such devastating scenarios help explain the numerous and diverse forest animals that became buried at the bottom of lakes, the seemingly charred corpses, and the virtual absence of microbial decomposition at the lakebed that fostered the exceptional preservation of the Jehol Biota. Undoubtedly, the rocks and the fossils make different types of volcanic-driven events accountable for the abundance and exceptional preservation of the Jehol Biota. Yet, how such scorching flows would have preserved many delicate soft structures in such an exquisite fashion, from tiny insect eggs to the finest details of feathers, remains unanswered.

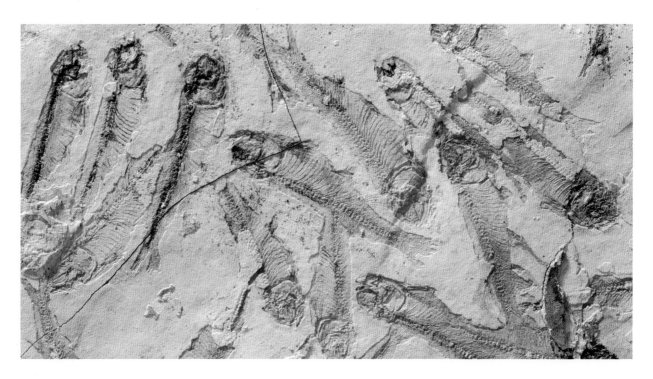

Confuciusornis sanctus
LPM-0228
Yixian Formation
Sihetun, Yixian County
Liaoning Province

Confuciusornis sanctus
LPM-0229
Yixian Formation
Sihetun, Yixian County
Liaoning Province

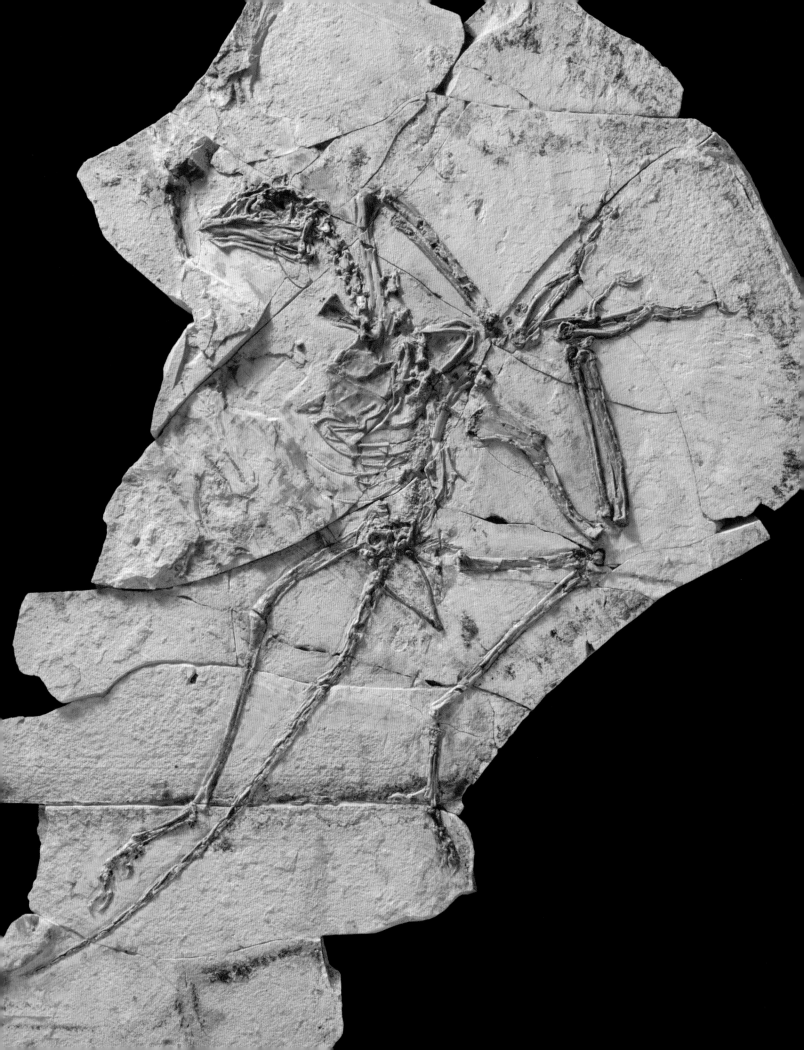

The Jehol fossils span a large portion of the family tree of early birds. Those that retained the long bony tail of *Archaeopteryx* and its dinosaurian predecessors represent the most primitive twigs of this tree and their fossils have been unearthed from various sites in Liaoning Province. Generally larger than most other Jehol birds, specimens of these long bony-tailed animals (*left* and *overleaf*) could have weighed up to 1.5 kilograms (slightly more than 3 pounds), roughly the size of a pheasant. The skeletons of these ancient birds look slightly more modern than that of the 150-million-year-old *Archaeopteryx*, even if their bony tail includes more vertebrae and is proportionally longer. The design of their feet and overall comparisons of their skeletons with those of living birds and mammals indicate that these Jehol birds spent much of their time on the ground feeding on grains and fruits, as revealed by examination of the gut contents preserved with their fossils.

Despite this, the anatomy of their skeletons and the characteristics of their plumage indicate a degree of aerial competence, although the nature and proficiency of their flight remains poorly understood. Several species of these primitive long-tailed birds have been named, but the differences used to separate them are likely due to bias in their preservation or simply to individual variation. So far, the best known among these species is *Jeholornis prima* and a smaller close relative, *Jeholornis palmapenis*. Yet, whether these two species are truly different remains controversial. One apparent difference between them is the presence in the latter of a palm-like arrangement of tail feathers (*below*).

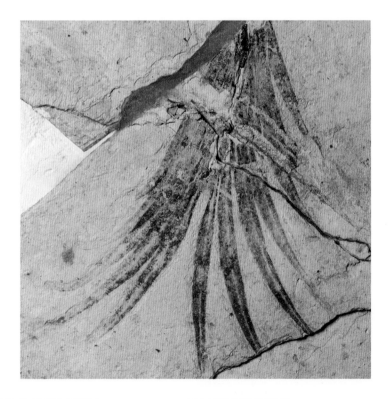

Jeholornis prima
BMNHC-PH780
Yixian Formation
Xiaoyugou, Chaoyang County
Liaoning Province

Jeholornis palmapenis
SDM-2000109
Jiufotang Formation
Jianchang, Chaoyang County
Liaoning Province

Jeholornis prima
CDL-02-04-001
Jiufotang Formation
Baitaigou, Yixian County
Liaoning Province

Complete and well-preserved fossils of *Jeholornis prima* and its kin have given us a clear idea of the anatomy and appearance of the long bony-tailed birds that 125 to 120 million years ago inhabited the Jehol forests. In many respects, these birds show a degree of modernization when compared to their older relative, *Archaeopteryx*. The short-snouted and tall skull of *Jeholornis* carried a few small, blunt teeth (*below*); these were limited to the front of the lower jaw and the back of the upper jaw. Details of the shoulder bones suggest that the Jehol long-tailed birds were capable of beating their wings over a larger arc than *Archaeopteryx*, boosting their ability to generate lift and thrust. The proportionally longer forelimbs and more compact hands of these birds (*left* and *previous spread*) also indicate a faster wing stroke, which was powered by muscles attached to a pair of large breastbones. These enlarged wings carried flight feathers that reached more than 20 centimeters (8 inches) in length.

Fossils of *Jeholornis prima* also reveal that a tuft of long shafted feathers projected from the end of this animal's extensive bony tail. These feathers might have played a role in flight control, helping to adjust pitch during takeoffs and landings, but it is also likely that they functioned as display traits. An example of the latter is the palm-like feathered tail of *Jeholornis palmapenis*, a design that suggests that early in the evolution of birds, feathers might have been already used in behaviors that today we see during mating contests and territorial claims.

Jeholornis prima
IVPP-V13274
Jiufotang Formation
Dapingfang, Chaoyang County
Liaoning Province

Jeholornis prima
BMNHC-PH780
Yixian Formation
Xiaoyugou, Chaoyang County
Liaoning Province

Living birds have a short bony tail that ends in a plate-like bone called the pygostyle, the result of the fusion of several vertebrae at the end of the spine. This bone provides support to a number of shafted tail feathers that generate extra lift and help birds navigate through the air. The abbreviation of the bony tail—its reduction in size and number of vertebrae—and the development of the pygostyle were clearly key evolutionary events in the history of birds, even if certain aspects of their precise function and evolution are not fully understood. Such reduction moved the center of gravity forward and led to new ways of both flying and walking, thus heralding the important diversification of birds that we see in the fossil record during the first half of the Cretaceous Period. How the extensive bony tail of primitive long bony-tailed birds, with its more than 20 long vertebrae, became shortened, and when this momentous transformation took place, are key questions that remain largely unanswered.

One unique Jehol fossil has given us a bit of clarity: a fledgling *Zhongornis haoae* (*left*), less than 10 centimeters (4 inches) long, in which the 13 to 14 short and unfused tail vertebrae fail to form a pygostyle. The anatomy of *Zhongornis* tells us that this tiny animal was possibly an intermediate form between long bony-tailed birds such as *Jeholornis* and more advanced ones with a reduced tail and a pygostyle. This immature specimen of *Zhongornis* thus suggests that the evolution of the avian bony tail might have passed through a stage in which the tail shortened by reducing the size and number of vertebrae before developing a pygostyle. Yet, not everybody agrees with the placement of *Zhongornis* as intermediate between long and short bony-tailed birds. A recent study has claimed that this unique fossil is in fact a member of the bizarre scansoriopterygids, a poorly known group of small non-avian theropod dinosaurs also known from the Jurassic-Cretaceous transition in northeastern China. However, this interpretation is controversial and by no means entirely convincing. Like many other paleontological debates, tipping the scale to one side or another will have to wait for new discoveries. For now, it seems that *Zhongornis* gives us a tiny answer to what are otherwise big open questions.

Zhongornis haoae
DNHM-D2456
Yixian Formation
Dawangzhangzi,
Lingyuan County
Liaoning Province

The abbreviation of the skeletal tail and the formation of the pygostyle, the bony stump at the end of the tail of living birds, were evolutionary events of great significance in the early history of these animals. In modern birds, the pygostyle provides support to a fatty protuberance—the rectricial bulbs, or more colloquially, the pope's nose—that anchor pairs (usually six) of long-shafted tail feathers. This structure also holds a suite of tiny muscles that control the movements of these feathers, and thus, the aerodynamic function of the feathered tail. The rectricial bulbs also support the preen gland, which produce an oily substance that birds distribute over their plumage and skin; it offers protection against exposure to sun, air, and water, and an antibiotic defense against parasites. The most primitive birds with a reduced bony tail and a pygostyle are known from the Jehol Biota. Whether these early short-tailed birds had already evolved the pope's nose or if this structure developed later has been hampered by the limited evidence of soft tissue associated with the tail. In fossil birds, the existence of this fundamental structure has been usually inferred by the presence of the shafted tail feathers it typically anchors.

A fossil (*left*) that lacks a set of long-shafted tail feathers suggests that this bird and many others in the Jehol Biota probably lacked a pope's nose. Such a conclusion is supported by ultraviolet visualization of some Jehol fossils, in which the bony tail does not appear to be surrounded by any large volume of soft tissue (*see previous pages*). This evidence suggests that the rectricial bulbs evolved after the reduction of the bony tail and the development of the pygostyle, although whether the earliest birds and their dinosaur forerunners also lacked a preen gland remains unclear. The precise functional advantage that came along with the evolutionary reduction of the bony tail remains somewhat elusive. Nonetheless, it seems clear that early in the evolutionary history of birds, the role the long skeletal tail and the feathers attached to it played in flight control was transferred to the wings of these animals. Such transition allowed the motion of the tail to be decoupled from the movement of the legs, fostering changes in the way these animals walked, ran, and flew. This momentous transformation made the wings largely responsible for the ability to fine-tune the precise adjustments that make birds masters of the air, and as such, it heralded the dramatic radiation these animals would experience during the first half of the Cretaceous Period.

Protopteryx fengningensis
BMNHC-PH1060A
Huajiying Formation
Sichakou, Fengning County
Hebei Province

In nature, some species are very common and others rare. Any given day in Los Angeles, one can watch dozens of American crows, rock pigeons, and house sparrows, and occasionally a scrub jay. Why some species are abundant and others are uncommon is not well understood, but the fossil record shows that such a pattern is as old as life itself. The crow-sized *Confuciusornis sanctus* is perhaps the most celebrated fossil from the Jehol Biota and definitively its most abundant bird. Hundreds of fossils of this species, from juveniles to adults, many with complete skeletons still surrounded by plumage (*left* and *overleaf*), have been unearthed from the Jehol rocks. *Confuciusornis* is also one of the most primitive short bony-tailed birds known worldwide. Its wings were long and tapering (*left*), resembling those of living terns, and a novel skinfold called the propatagium formed a triangular extension of the wing that connected the wrist to the shoulder. The end of its short bony tail bore a large pygostyle (*below*) that was covered by fluffy feathers; in some individuals (*left*), this tail stump anchored a pair of extremely long, ornamental feathers.

These and other aspects of this ancient animal indicate that even the most primitive short-tailed birds were likely better fliers than their forerunners with long bony tails. The unparalleled fossil collection of *Confuciusornis sanctus* also allows us to study the individual development and physical variation of one of the earliest known bird species; the sheer numbers of *Confuciusornis* enable us to analyze a wealth of details about this archaic bird using essentially the same statistical tools ornithologists use when studying modern-day species.

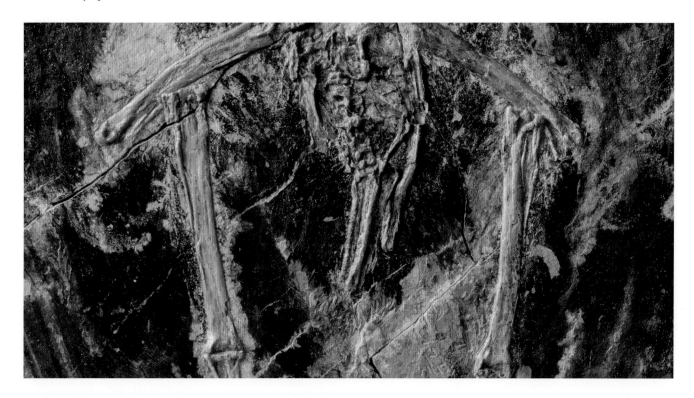

Confuciusornis sanctus
DNHM-D2859
Yixian Formation
Sihetun, Yixian County
Liaoning Province

Confuciusornis sanctus
BMNHC-PH987
Yixian Formation
Sihetun, Yixian County
Liaoning Province

Confuciusornis sanctus
BMNHC-PH766
Yixian Formation
Sihetun, Yixian County
Liaoning Province

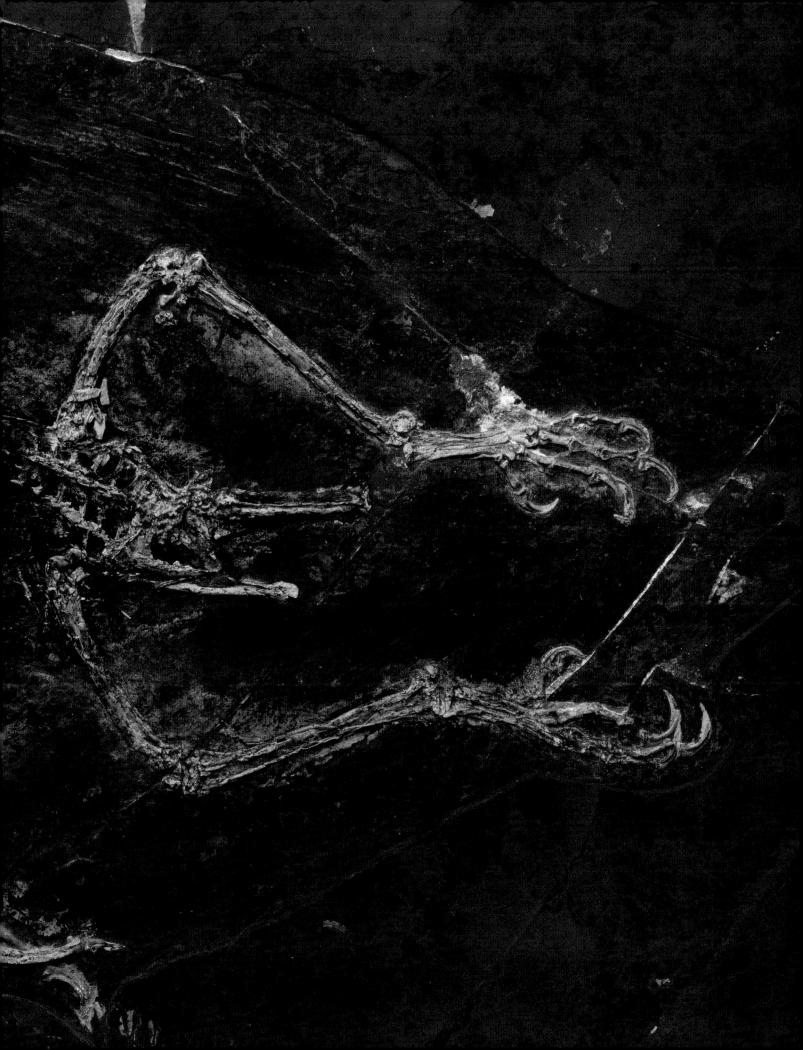

One of the most distinctive features of *Confuciusornis* is its strong, massive, and toothless beak (*below*), which seems ideally suited for cracking tough seeds and opening hard fruits. In fact, *Confuciusornis* is the most primitive example of a beaked bird. However, while superficially similar to all living birds in this respect, detailed examination of the skull of this 125-million-year-old bird reveals many primitive features that attest to its archaic nature. Another characteristic feature of this ancient bird is the stout appearance of the wing bones and a hand that ended in long fingers with sharply curved claws (*left* and *previous spread*) that most likely projected from the edge of the wing's plumage. The legs of this bird were strong and covered with soft plumage, which extended down to the ankles. The foot had a grasping appearance, but the short opposable hind toe tells us that *Confuciusornis* was not a specialized percher.

The toothless beak, the powerful claws of its hands, and the unspecialized feet are all recognizable features of *Confuciusornis sanctus*, but what often stands out in many of its fossils is the enormously long pair of ornamental feathers that extended from the tail (*overleaf*). Interestingly, not all specimens of this bird have them. Close examination of those individuals without these long and stiff feathers indicates that their absence is not due to a lapse of preservation in the neighboring rock. The fact that some of these fossils display these long tail feathers while others do not has been regarded as an indication that the feathers are a sexual trait, their presence documenting an ancient courtship and allowing us to distinguish the ornate male from the drab female.

Confuciusornis sanctus
BMNHC-PH931
Yixian Formation
Sihetun, Yixian County
Liaoning Province

Confuciusornis sanctus
BMNHC-PH986
Yixian Formation
Sihetun, Yixian County
Liaoning Province

Confuciusornis sanctus
BMNHC-PH987
Yixian Formation
Sihetun, Yixian County
Liaoning Province

BMNHC
Ph000987

Differentiating one species from another in the fossil record—particularly if they are close relatives—is never easy and almost always controversial. Over the years, several species of *Confuciusornis* have been named but, as in the case of other Jehol birds, this has often been the result of reading too much into small differences that are common in natural populations or frequently the result of fossil preservation. One example of splitting such differences too thin is *Confuciusornis feducciai*, a species based on a crow-sized specimen (*right*) exhibiting subtle variations in the shape and proportions of several bones. Yet, when placed in the context of the large known sample of *Confuciusornis sanctus*, these differences seem within the normal range of avian populations. Fortunately, the extraordinary number of specimens of *Confuciusornis sanctus* from various sites in Liaoning Province has given us the opportunity to understand the physical variation of these birds from a statistical perspective.

Measurements of the bones of hundreds of skeletons of *Confuciusornis sanctus* give us a wealth of information about how the skeletons of these birds varied from one another: in the transition from juvenile to adult, in the differences between males and females, and in the way that not all individuals of a species are identical to one another. When all this information is analyzed together with that of the different species that have been named, it shows that the specimens used to name other species of *Confuciusornis* are likely members of the same species, *Confuciusornis sanctus*. The differences between these specimens are either the result of subtle variations in skeletal proportions typical of any natural population, modifications that occurred as animals grew and became adults, or simply differences in how their bones were preserved, which is a significant variable in Jehol fossils that are largely preserved in two dimensions. The subtle differences observed among the hundreds of fossils of *Confuciusornis sanctus*, which statistically are revealed as belonging to a single species, give us a cautionary note about reading too much into the skeletons of animals that, while beautifully preserved, are still flattened against a slab of rock.

Confuciusornis sanctus
DNHM-D2454
Yixian Formation
Sihetun, Yixian County
Liaoning Province

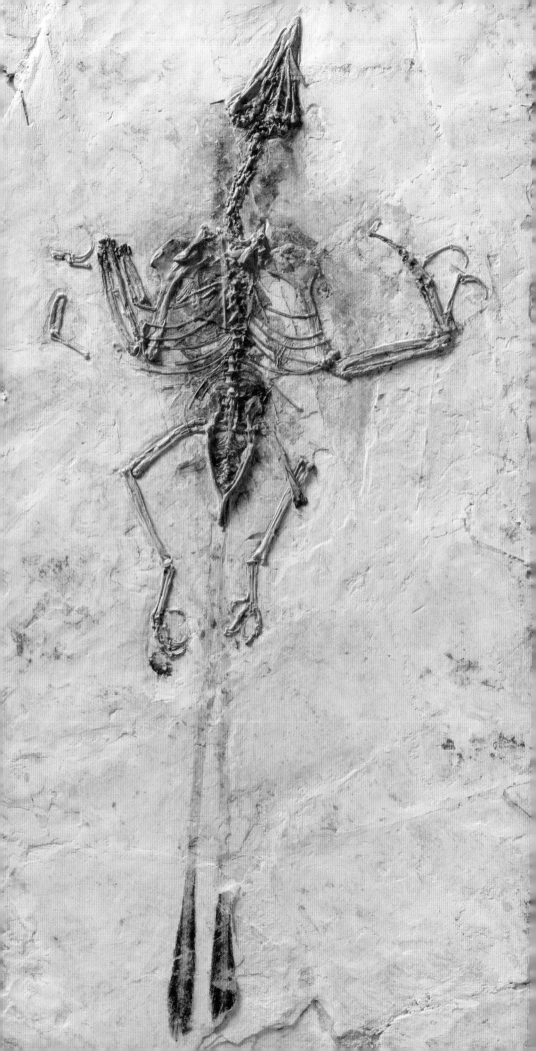

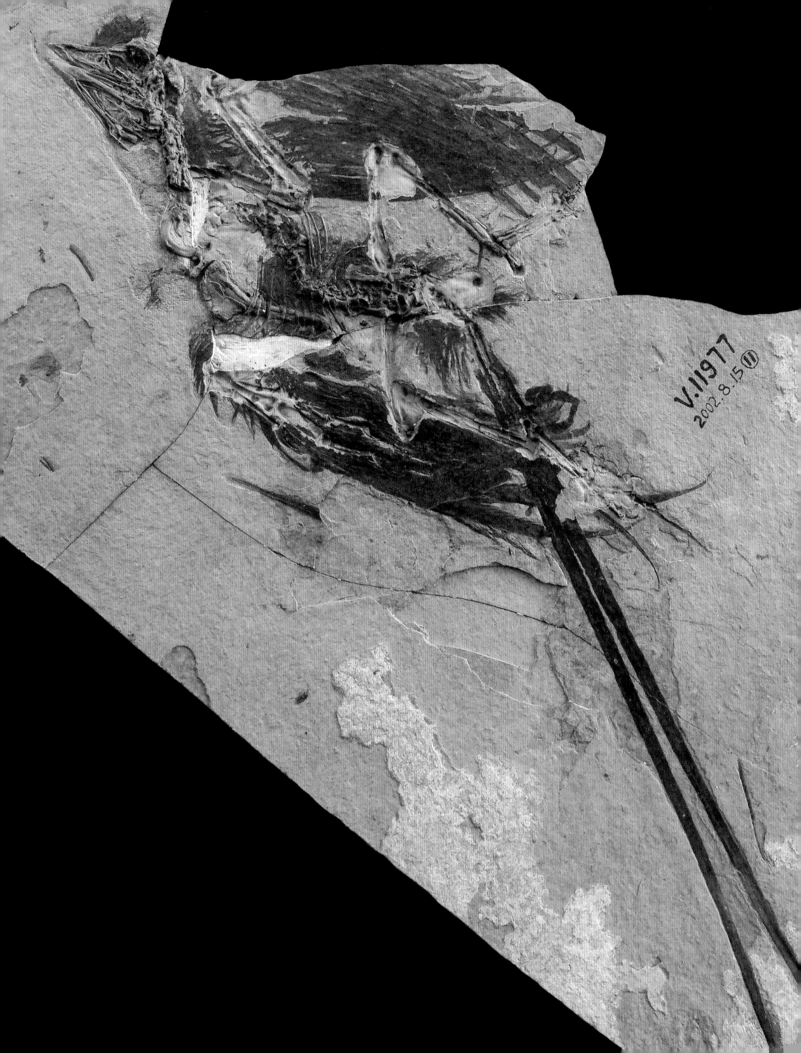

Discoveries of close relatives of *Confuciusornis sanctus* living both at the same time and millions of years earlier are providing information about this fascinating bird group. The oldest of them is *Eoconfuciusornis zhengi*, which lived 131 million years ago in what is today northern Hebei Province. We know of just a handful of fossils of this ancient bird and most of our knowledge comes from an immature specimen with exquisite plumage (*left*) that is similar in most respects to *Confuciusornis sanctus*. Like many fossils of the latter, this specimen carries a pair of ornamental tail feathers that reveals that the males of *Eoconfuciusornis* were more ornate than the females. Such gaudy plumage in a 131-million-year-old fossil gives us one of the earliest known examples of differences between the plumage of males and that of females in birds. The fact that these feathers are present in a juvenile also indicates that these early birds began reproducing before they were full-grown, the reverse of what commonly happens among today's birds.

Another cousin of *Confuciusornis*, but most likely less closely related to it than *Eoconfuciusornis*, is *Changchengornis hengdaoziensis* (*below*). Smaller and lighter than *Confuciusornis*, *Changchengornis* lived at the same time and in the same forest environments that hosted flocks of its cousin. Its beak was shorter, more delicate, and more downcurved than the powerful beak of *Confuciusornis*. These features suggest that while sharing the same forests, these two birds relied on different food sources. The discovery of these two poorly known relatives of *Confuciusornis*—*Eoconfuciusornis* and *Changchengornis*—underscores the existence of a hidden diversity of these beaked birds waiting to be unearthed.

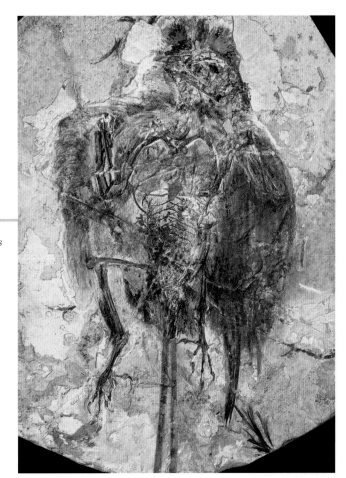

Changchengornis hengdaoziensis
GMV-2129B
Yixian Formation
Jianshangou, Yixian County
Liaoning Province

Eoconfuciusornis zhengi
IVPP-V11977
Huajiying Formation
Sichakou, Fengning County
Hebei Province

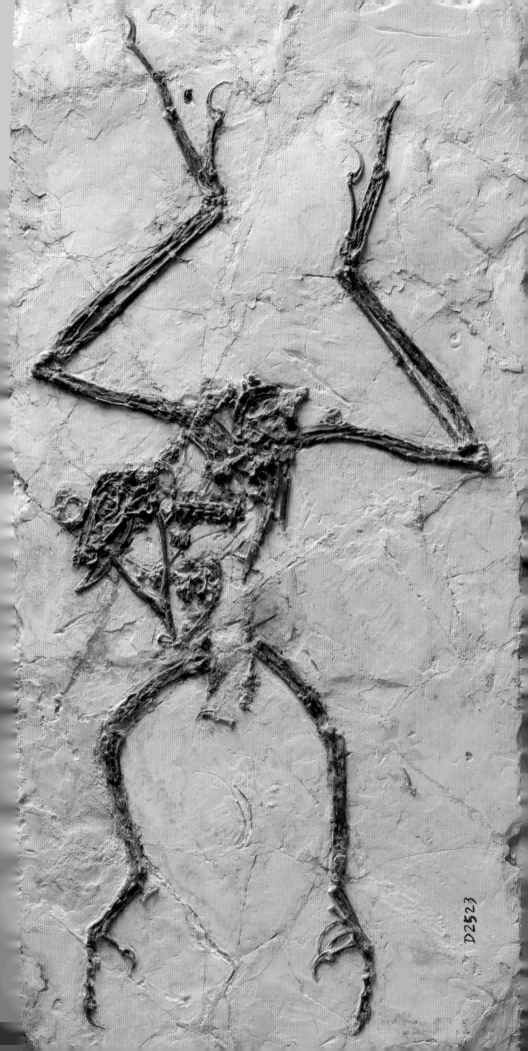

D2523

Known exclusively from the Jehol Biota and from sites ranging between 125 and 120 million years in age across Liaoning Province, *Sapeornis chaoyangensis* (*left* and *overleaf*) wrangles with *Confuciusornis* and its kin for the title of most primitive short bony-tailed bird. Reaching up to 1.2 kilograms (a little more than 2.5 pounds), roughly the weight of a western gull, this spectacular animal is the largest short bony-tailed bird known from the Jehol and elsewhere during the first half of the Cretaceous Period. Much longer than the legs, the formidable arms of *Sapeornis* supported long and relatively broad wings that gave this bird an estimated 110-centimeter (43 inches) wingspan. Its wing feathers had an essentially modern appearance. Even the number of primaries and secondaries—the flight feathers attached to the hand and forearm, respectively—was comparable to those typical of living birds.

Fossils of *Sapeornis* tell us of an important but yet obscure chapter of the early history of avian flight. The magnificent wings of these birds were accompanied by a very primitive design of the shoulder, which together with the lack of breastbones, have made researchers wonder about the aerodynamic competence of these animals. Their lifestyle is also poorly understood, although available evidence from their skeletons as well as gut contents suggests that they foraged on the ground, eating seeds and fruits. In contrast, its powerfully clawed feet, with a relatively long and opposable hind toe and large, coarsened pads protecting the undersides of the toes (*below*), resemble the feet of today's birds of prey, and make us wonder whether *Sapeornis* used its grasping feet to subdue smaller prey.

Sapeornis chaoyangensis
DNHM-D2523
Jiufotang Formation
Gonggao, Chaoyang County
Liaoning Province

Sapeornis chaoyangensis
HGM-41H1110405
Jiufotang Formation
Dapingfang, Chaoyang County
Liaoning Province

Although much less common than *Confuciusornis*, fossils of *Sapeornis chaoyangensis* give us a good idea of the overall appearance of this ancient bird. Its deep skull, at the end of a relatively long neck, bore 14 to 16 spade-shaped teeth on the combined upper jaws (*right*). In contrast, the lower jaws were toothless, and horny sheaths might have covered them. The function of the teeth, with no counterpart to abut against, is unclear, especially when known fossils document that these birds had an essentially modern digestive system in which food was processed by a highly modified stomach that was as efficient at breaking up food as the elaborate teeth of mammals. The tremendously long arms of *Sapeornis* ended in sharp claws at the tips of the innermost and middle fingers (*see previous spread*), which were likely visible through the wing's plumage. These wings must have been powered by large flight muscles that attached either to cartilaginous breastplates (the breastbones of these birds have never been found) or to expansive areas of some of their shoulder bones. The abbreviated bony tails of these birds ended in a short pygostyle, the plate-like bone at the end of the tails of modern birds. While the plumage of the tail is not well known, some specimens (*below*) hint at the presence of long feathers that attached to their short bony tails. Known fossils also show that feathered trousers covered the legs, although the extent of the leg plumage is not entirely clear.

Sapeornis chaoyangensis
BMNHC-PH1067
Yixian or Jiufotang Formation
Linglongta, Jianchang County
Liaoning Province

As with other Jehol birds, the species diversity of *Sapeornis* might have been overstated. *Sapeornis angustis*, for example, was recognized on the basis of a specimen (*right*) in which proportions of several long bones differ from those of larger specimens of *Sapeornis chaoyangensis*. Two-thirds of the size of a full-grown *Sapeornis chaoyangensis*— a bird as large as a western gull—the specimen used to name *Sapeornis angustis* shows immature features, indicating it died while it was still young. Not surprisingly, statistical analyses of the known fossils of *Sapeornis* show that the differential bone proportions that supported the erect posture of *Sapeornis angustis* are developmental; they correspond to changes that took place as juveniles grew into adults.

Another example of taxonomic over-splitting involves the fossil used to name the species *Didactylornis jii* (*below*). In this case, detailed examination shows that the anatomy of the hand was misinterpreted—the innermost hand bone being incorrectly identified as part of the innermost finger—and that other perceived differences are the result of poor preservation. In the end, when the specimens used to create these and another named species of *Sapeornis* are studied in the context of a large number of other fossils, all of them appear be indistinguishable from the originally named species, *Sapeornis chaoyangensis*.

Sapeornis chaoyangensis
CDL-08-02-01
Jiufotang Formation
Baitaigou, Yixian County
Liaoning Province

Sapeornis chaoyangensis
IVPP-V13396
Jiufotang Formation
Dapingfang, Chaoyang County
Liaoning Province

Birds that live in forests, particularly those surrounding rivers and lakes, tend to show a greater diversity than those inhabiting other natural habitats. Trees offer food, protection, and nesting areas. Not surprisingly, the Jehol enantiornithines (*left*)—birds that are consistently interpreted as arboreal—are represented by a larger number of species than those known for other groups of birds from the same sites. In fact, the enantiornithines were the most species-rich birds of the Mesozoic Era. Their fossilized remains are found on every continent (except Antarctica) and in environments ranging from inland deserts to rivers, lakes, and marine coastal settings. Their abundance and variety, and the different environments they occupied, illustrate their great success and broad ecological diversity. Remains of enantiornithines are indeed quite common in the Jehol Biota, their beautiful fossils often surrounded by plumage (*overleaf*).

During the past three decades, thousands of fossils representing more than 30 enantiornithine species have been unearthed from quarries in Liaoning, Hebei, and Inner Mongolia. Sites in northern Hebei Province also document the world's earliest appearance of these birds, which dates to about 131 million years ago. Even at this early juncture in their history, the fossils of these birds show many of the trademarks of the group. Millions of years later, the successful saga of the enantiornithines would end abruptly, as these birds disappeared leaving no descendants during the deadly extinction that some 66 million years ago also wiped out the last of their non-avian dinosaur forerunners. Numerous aspects of the skeletons and plumage of the enantiornithines—exquisitely preserved in many Jehol fossils—document a significant improvement in the flight capabilities of these birds when compared to those of their more primitive counterparts. The Jehol enantiornithines were typically songbird-sized, and had a range of wing proportions that heralds a diversity of flight styles and performance never reached by their forerunners. The superior flying abilities of the earliest enantiornithines must have paved the road for their future dominance and for the diversity of lifestyles, varieties, and sizes these birds would evolve during their vast evolutionary history.

Sulcavis geeorum
BMNHC-PH805
Yixian Formation
Lamadong, Jianchang County
Liaoning Province

Enantiornithine indeterminate
DNHM-D2884/1
Yixian or Jiufotang
Formation
Liaoning Province

When compared to the skeletons of more primitive birds, those of enantiornithines exhibit an interesting combination of features: some characteristics have remained largely unchanged, while others have been highly modified. The shape of the skull, for example, is in many ways similar to that of *Archaeopteryx* and its dinosaurian forerunners (*below*). The wing bones, however, show a notable degree of modernization—the fingers of enantiornithines became shorter, with the outermost one lacking a claw, and the proportions of the main bones grew akin to those of modern-day birds (*left*). The shoulder bones also acquired a more modern appearance. The enantiornithine breast-bone became proportionally larger and better suited for accommodating massive flight muscles, and the wishbone evolved into a slender Y-shaped structure more similar to the wishbone of many living birds that assists the wing beat during flight. This interesting mosaic of primitive skulls and more modern-looking wings and shoulder bones reveals a different pace in the evolution of these body parts, with an emphasis on structures key to flight performance.

The foot of enantiornithines, particularly among the Jehol species, shows a distinctly grasping appearance. The long hind toe points opposite to the other toes and often carries a large claw. This design is closer to the refined perching foot of many songbirds and other present-day birds that live in bushes and trees. Coupled with the small sizes typical of most Jehol enantiornithines, the grasping appearance of their feet indicates that most Jehol enantiornithines called the lush forests of this ancient world their home.

Shenqiornis mengi
DNHM-D2950
Qiaotou Formation
Senjitu, Fengning County
Hebei Province

Protopteryx fengningensis
BMNHC-PH1060A
Huajiying Formation
Sichakou, Fengning County
Hebei Province

About 131 million years ago, in what is today northern Hebei Province, the enantiornithines made their debut in the fossil record. Heralding their success for the rest of the Mesozoic Era is the starling-sized *Protopteryx fengningensis* (*left* and *overleaf*), perhaps the most primitive known enantiornithine bird. The appearance of *Protopteryx* in the fossil record is abrupt; this most ancient enantiornithine already displays many of the characteristics typical of the group. While toothed and equipped with sharp claws at the ends of its fingers, even this early enantiornithine shows important modifications in the bones that provide structure to the feathered wings and flight muscles, and a considerable reduction in size, which speak of a significant evolutionary redesign with respect to its larger forerunners.

A handful of exquisitely preserved fossils of this bird—all from the earliest stages of the Jehol Biota—give us a good understanding of its appearance. Its wings were long and tapered; its skull had a short snout that sported small, sharp teeth on both upper and lower jaws; its feet carried strong claws and toes designed for grasping (*below*); and fluffy, down-like feathers covered its body. The presence of a pair of long ornamental feathers decorating the tail of this animal (*overleaf*) once again attests to the important role played by the plumage during courtship, even at the earliest phases of bird evolution. The relatively small size and perching feet of *Protopteryx* suggest that this bird was already well adapted for life in trees, presumably living off the abundant insects that inhabited the Jehol forests.

Protopteryx fengningensis
BMNHC-PH1158A
Huajiying Formation
Sichakou, Fengning County
Hebei Province

Protopteryx fengningensis
BMNHC-PH1060A
Huajiying Formation
Sichakou, Fengning County
Hebei Province

Protopteryx fengningensis
BMNHC-PH1060A\B
Huajiying Formation
Sichakou, Fengning County
Hebei Province

Size plays a major role in the lives of animals—it defines interactions with both the environment and other organisms and constrains many biological properties, including physiology, reproduction, and locomotion. The evolutionary transition from large dinosaurs to birds involved a dramatic miniaturization in which the forerunners of living birds were downsized as the result of a prolonged trend that started more than 50 million years prior to the appearance of the earliest birds. With sizes that in many cases fall within the low range of today's songbirds, enantiornithines were the smallest known birds of the Mesozoic Era and a group that shows great variation in size. Indeed, during their 65-million-year evolutionary history, enantiornithines evolved dimensions that ranged between those of the smallest songbirds and those of vultures. While the spectrum of enantiornithine size was narrower among the Jehol species, the different sizes of these early birds are obvious when comparing fossils of adult individuals. For example, the femur (thigh bone) of the largest known specimen of *Pengornis houi* (*left*), the biggest enantiornithine in the Jehol Biota, is close to two-and-a-half times that of the same bone from another, smaller enantiornithine that could easily fit in the palm of a hand (*overleaf*)—these two fossils are here shown at their actual size.

Estimates of the size and weight of extinct animals are typically based on mathematical extrapolations from the dimensions of the bones of their living relatives, and, not surprisingly, these estimates differ significantly when they are based on different samples of living animals or different sets of bones. Thus, knowing the real weight of these ancient birds is always difficult. *Pengornis* could have tipped the scale at about 400 grams (slightly less than 1 pound), close to the size of an American crow, while the tiny enantiornithine featured in the following pages would have been an order of magnitude smaller, hardly weighing more than 40 grams (1.5 ounces). The size range exhibited by the Jehol enantiornithines is clearly a reflection of the specialized lifestyles of these birds, which undoubtedly used different resources and exploited their environments differently. Such size variation also points to differences in how these animals flew because, among other things, size is a determining factor in the aerodynamics of flying animals.

Pengornis houi
IVPP-V15336
Jiufotang Formation
Dapingfang, Chaoyang County
Liaoning Province

Enantiornithine indeterminate
BMNHC-PH877
Yixian Formation
Jianchang County
Liaoning Province

From the straight and delicate bill of a hummingbird to the massive and curved beak of a flamingo, modern birds display an enormous variation in the shape of their skulls, which is directly related to what these animals eat. The skulls of the Jehol enantiornithines also show a great deal of variation, reflecting the different feeding and ecological specializations of these animals. Just as in modern birds, the greatest range of this diversity lays in the shape and size of the snout, which plays a key role in how birds procure their food and build their nests. Some enantiornithines had elongated skulls with delicate snouts used for probing soft substrates. In others, the long snout sported large teeth (*below*), necessary for securing slippery fish. Many had short snouts with pointy teeth (*left* and *overleaf*) that might have been ideal for grabbing insects.

Detailed examination of their skulls reveals a notable degree of conservatism that goes way beyond the presence of teeth. The skulls of the Jehol enantiornithines are in many ways similar to the skull of *Archaeopteryx* and some of its dinosaurian forerunners, and quite different from those of their modern relatives. These primitive features impacted not only the motion of the jaws but also the size and structure of the brain. The advanced flight features of even the earliest enantiornithines tell us of a significant improvement in the aerial capabilities of these animals when compared to that of *Archaeopteryx* and other most primitive birds, yet the archaic appearance of their skull suggests that the cranial and sensory evolution of the enantiornithines lagged behind that of portions of the body involved in the mechanics of flight.

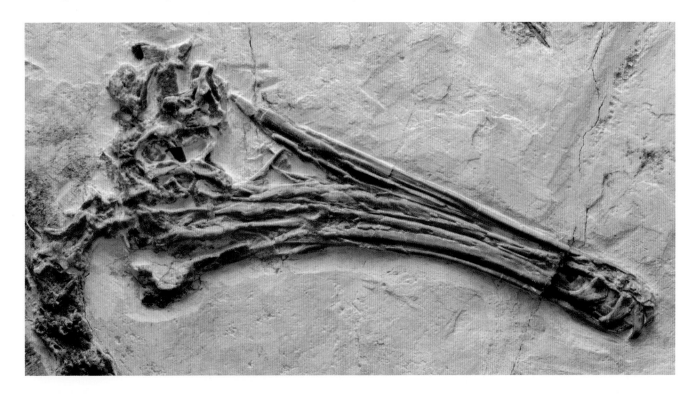

Bohaiornithidae
BMNHC-PH1204
Jiufotang Formation
Lamadong, Jianchang County
Liaoning Province

Longipteryx chaoyangensis
DNHM-D2889
Yixian Formation
Yuanjiawa, Chaoyang County
Liaoning Province

Eoenantiornis buhleri
IVPP-V11537
Yixian Formation
Heitizigou, Chaoyang County
Liaoning Province

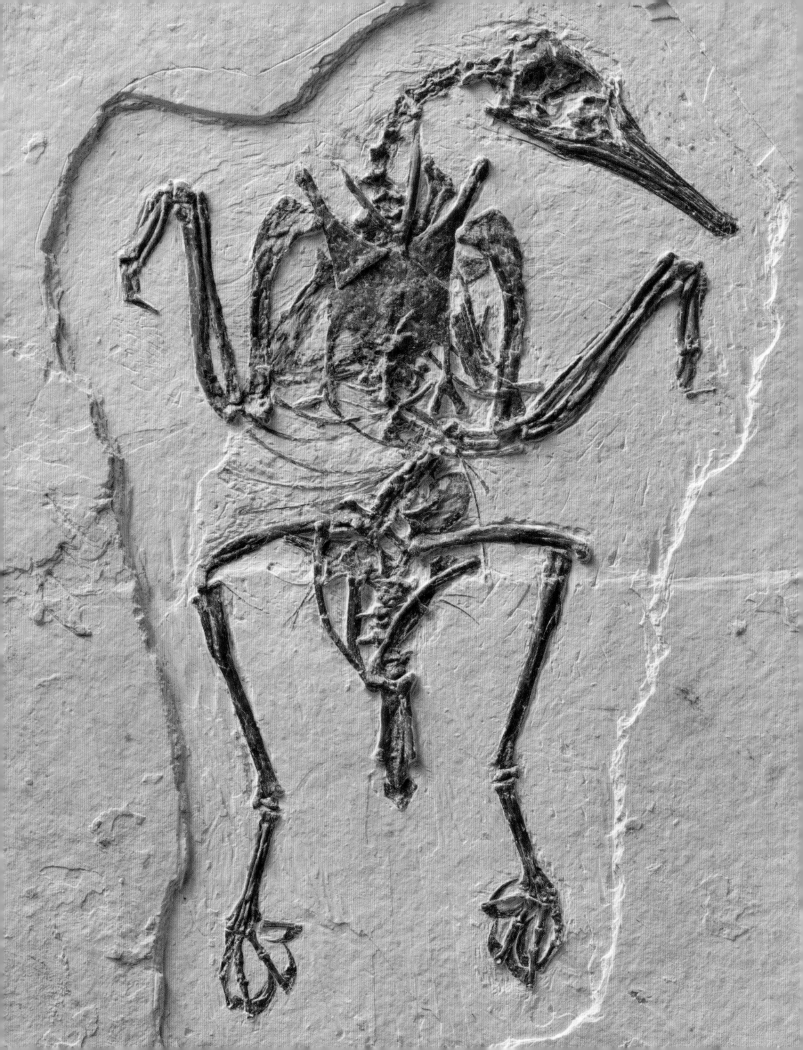

Most of the birds that lived during the Mesozoic Era had teeth set in deep sockets. The vast majority of enantiornithines, including all known species from the Jehol Biota, were toothed as well. The teeth of these birds varied greatly in shape, size, number, and placement in the jaws. *Sulcavis geeorum* (*left*) might have carried up to 40 teeth distributed over much of its jaws; its stout teeth appear suited for cracking hard items. The teeth of the Jehol enantiornithines were typically high and pointy, but *Pengornis houi* (*overleaf*) had teeth with low and dome-shaped crowns. *Longipteryx chaoyangensis* (*below* and *overleaf*) had a handful of large, recurved teeth restricted to the tips of the jaws. In some of its relatives, *Rapaxavis pani* and *Longirostravis hani*, among others, the teeth were also clustered at the front of the jaws but they were tiny and peg-shaped.

Such a remarkable range in tooth shape and position underscores the wide variety of diets and feeding specializations of these birds. The delicate, pointy teeth of many of them might have been effective for piercing the soft bodies of the numerous insects that inhabited the ancient Jehol environments. The low-crowned teeth of *Pengornis* seem to be well adapted for crushing relatively hard food items, and the stout but pointy dentition of *Sulcavis* might have been suited for cracking harder foods, such as the shells of mollusks. The large teeth at the tips of the jaws of *Longipteryx* most likely helped in holding on to slick fish; in some fossils these teeth have crenulations on the rear edges of the crowns that resemble the saw-like edge of a harpoon (*below*).

(Previous spread)

(left page)
Rapaxavis pani
DNHM-D2522
Jiufotang Formation
Lianhe, Chaoyang County
Liaoning Province

(right page)
Zhouornis hani
BMNHC-PH756
Jiufotang Formation
Xiaoyugou, Chaoyang County
Liaoning Province

83

Sulcavis geeorum
BMNHC-PH805
Yixian Formation
Lamadong, Jianchang County
Liaoning Province

Longipteryx chaoyangensis
DNHM-D2889
Yixian Formation
Yuanjiawa, Chaoyang County
Liaoning Province

Longipteryx chaoyangensis
BMNHC-PH826
Jiufotang Formation
Dapingfang, Chaoyang County
Liaoning Province

Pengornis houi
IVPP-V15336
Jiufotang Formation
Dapingfang, Chaoyang County
Liaoning Province

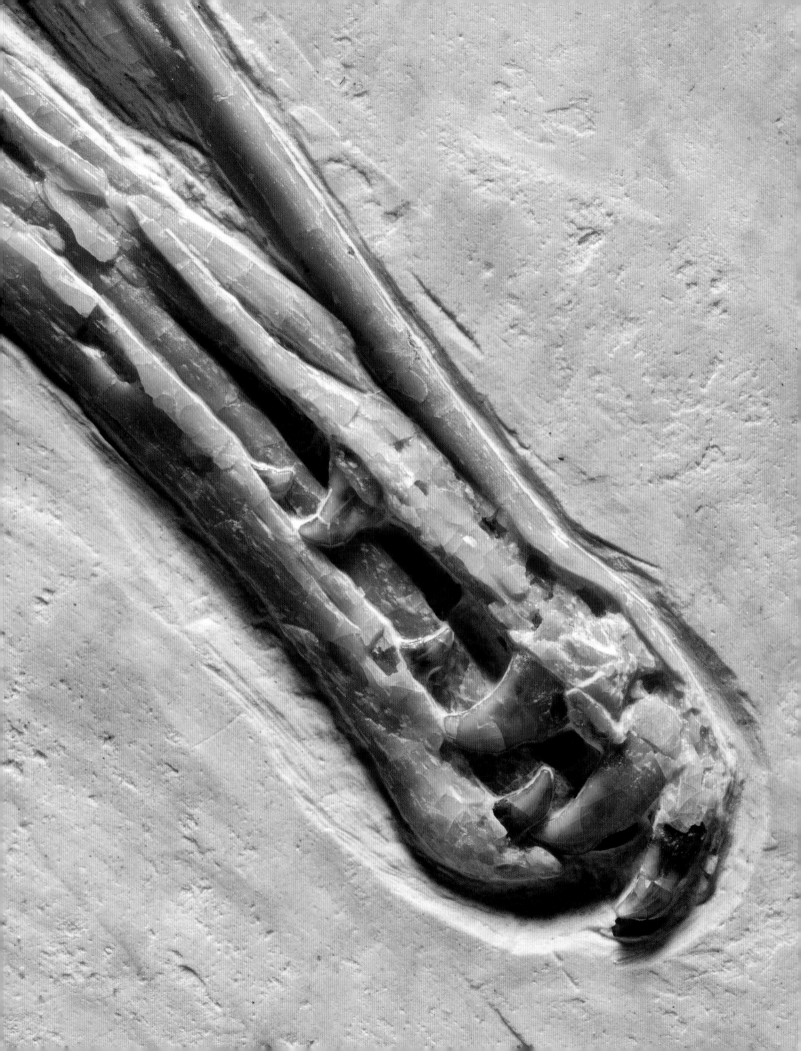

The many discoveries of early birds from the Jehol Biota have led to a much greater understanding of the evolutionary relationships within the family tree of enantiornithines. The exceptional fossil record of enantiornithines from these sites has revealed a number of well-defined groups with their own anatomical characteristics and specialized lifestyles. One of these distinct groups includes a diversity of small species with long and delicate snouts that at their ends carried a handful of diminutive, peg-like teeth (*left* and *below*). *Longirostravis hani*, *Rapaxavis pani*, and others are all part of this family of small enantiornithines, roughly the size of a western sandpiper and known exclusively from the Jehol Biota. The stratigraphic distribution of these animals indicates that as a group they lived for at least 5 million years, from approximately 125 to 120 million years ago. Their long and delicate snouts might have been ideal for probing the mud and soft surfaces along the coasts of the Jehol lakes.

While fossils of *Longirostravis* and its kin are exceedingly rare, each species known by one or a few specimens, a purported relative of these birds—*Longipteryx chaoyangensis*—is among the most common birds in the Jehol Biota (*overleaf*). With a size comparable to that of a pied kingfisher, *Longipteryx* is substantially larger than its more delicate cousins and much more stoutly built. At the end of its long snout, it carried a handful of teeth that, unlike those of *Longirostravis* and kin, were recurved, compressed, and massive; the long jaws and teeth designed for securing slippery prey indicate that *Longipteryx* was a fish-eater.

Longirostravis hani
IVPP-V11309
Yixian Formation
Yixian County
Liaoning Province

Longipteryx chaoyangensis
HGM-41HIII0319
Yixian or Jiufotang
Formation
Liaoning Province

Longipteryx chaoyangensis
BMNHC-PH1071
Jiufotang Formation
Lamadong, Jianchang County
Liaoning Province

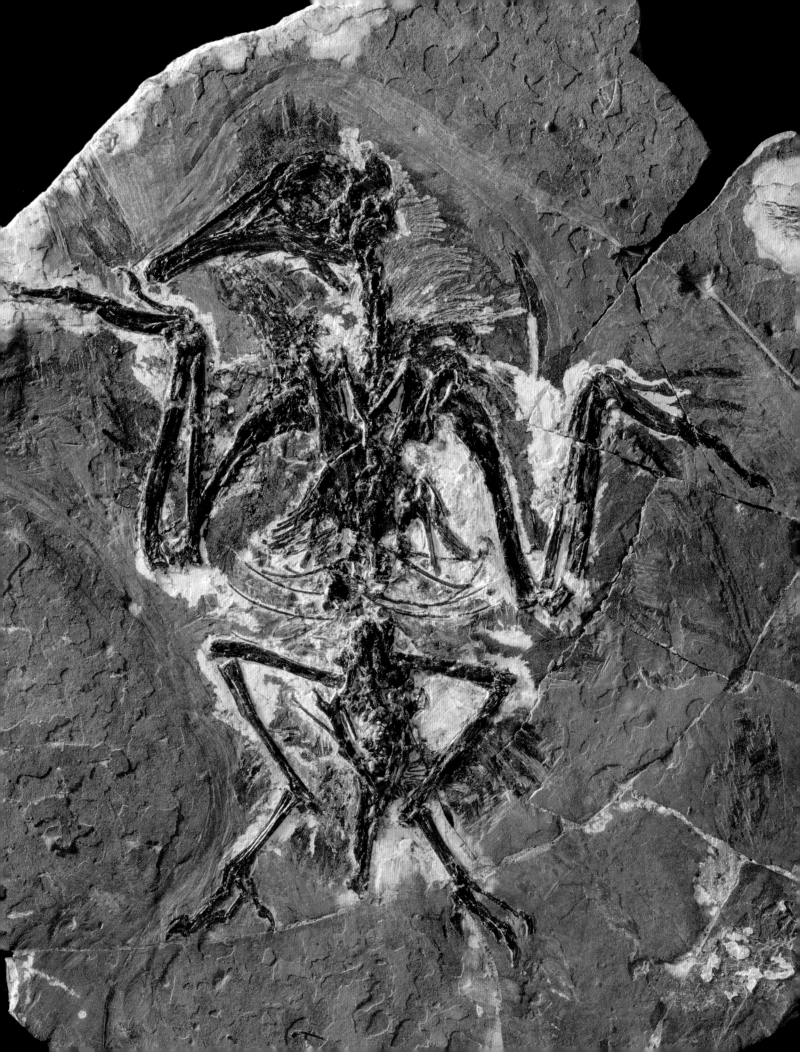

Another recently recognized group of enantiornithines, the bohaiornithids (*left*), includes more than a half-dozen species of pigeon-sized birds distinguished by subtle differences in their skeletons. Known exclusively from the Jehol Biota, and from rocks dating between 125 and 120 million years ago, these birds are relatively easily identified when compared to other enantiornithines. They had strongly built skulls and tough, pointy teeth that were evenly distributed on both the upper and lower jaws (*overleaf*). Their arms were slightly shorter than their stout legs, which had short feet with long and gently curved claws (*below*). The plumage of these birds is not well preserved in any of the known fossils, but their wings seem to have had an elliptical outline. The broad hip and long thighbone of these animals (*left*) suggest they spent a good amount of their time foraging on the ground.

Larger than those of most other Jehol enantiornithines, the feet of these birds have been interpreted as raptorial. However, whether these birds were predators and, if so, what they typically preyed on, is difficult to say because no fossil of Jehol enantiornithines (including all bohaiornithids) has preserved any evidence of gut contents. The uniqueness of the skeletons of the bohaiornithids speaks of a specialized mode of life, but we may need to wait for future discoveries to be able to better understand the lifestyles of these animals.

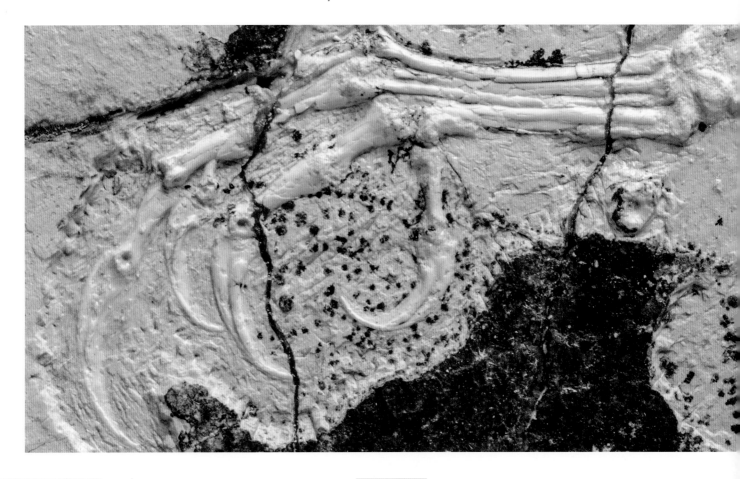

Bohaiornithidae
BMNHC-PH1204
Jiufotang Formation
Lamadong, Jianchang County
Liaoning Province

Zhouornis hani
BMNHC-PH756
Jiufotang Formation
Xiaoyugou, Chaoyang County
Liaoning Province

Zhouornis hani
CNUVB-0903
Jiufotang Formation
Chaoyang County
Liaoning Province

The enantiornithines were the most species-rich group of birds in the Jehol Biota, but these animals shared their world with a menagerie of other birds (*left*) that depict the initial phases of the evolution of the group that includes all living birds. Fossils from sites in Liaoning and Hebei Provinces show that a number of these primitive ornithuromorphs, as the group is called, inhabited the Jehol's ancient habitats. The earliest of them, the 131-million-year-old *Archaeornithura meemannae* from Hebei's Huajiying Formation, is the oldest known ornithuromorph worldwide. Many fossilized remains of other ornithuromorphs are known from the younger Yixian and Jiufotang Formations in Liaoning Province. The anatomy of each of these lineages progressively approaches that of their living relatives, but even the earliest of them show a significant degree of modernization in portions of the skeletons involved with both flying and landing.

These fossils present us with the clearest window into the appearance, behavior, and lifestyle of the earliest forerunners of living birds; they also document how many of the features of today's birds evolved more than 100 million years ago. The anatomy of the Jehol ornithuromorphs hints at important ecological differences with respect to their enantiornithine contemporaries. Their generally larger size and lack of perching adaptations (*below*) greatly contrasts with the small size and perching feet typical of the Jehol enantiornithines. Overall, early ornithuromorphs show features that are better adapted for either a terrestrial or a semiaquatic lifestyle. Their fossils tell us that the most primitive relatives of living birds lived either on the forest floor or in the coastal regions of ancient lakescapes.

Yanornis martini
XHPM-1205
Yixian Formation
Jinzhou, Yixian County
Liaoning Province

Schizooura lii
IVPP-V16861
Jiufotang Formation
Jianchang County
Liaoning Province

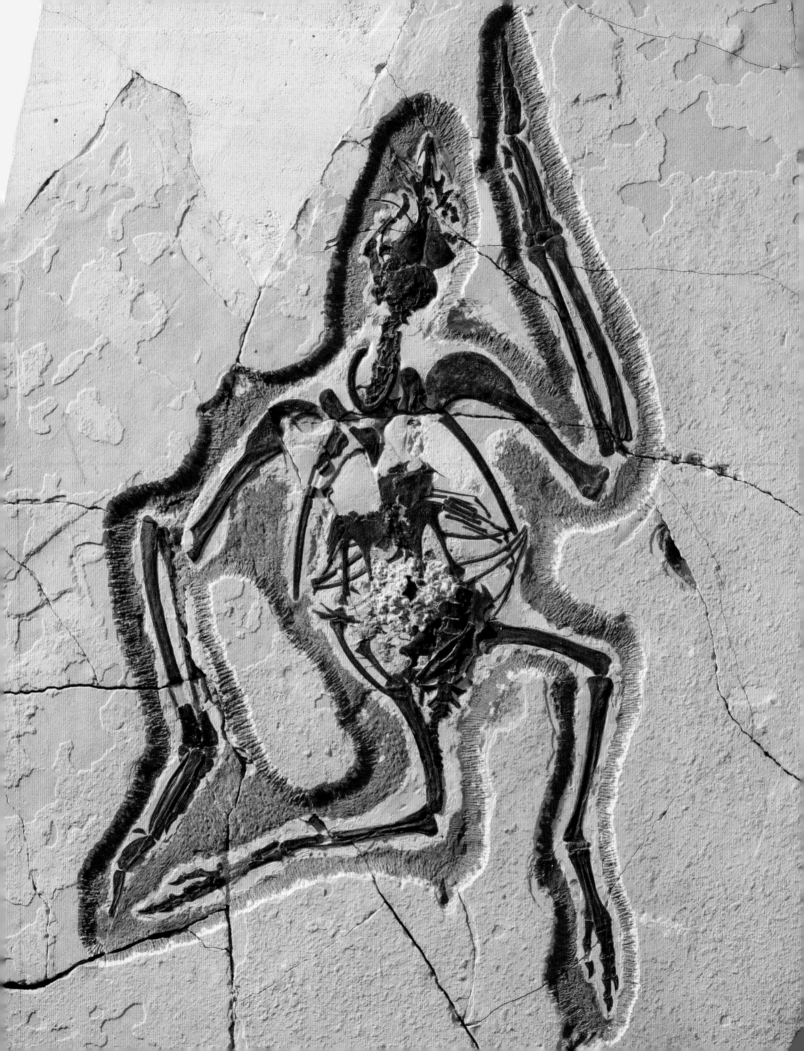

Transitional forms, those intermediate between well-defined groups of organisms, are often rare. In this respect, the Jehol Biota provides another important contribution to our understanding of the early evolution of birds. Some of these birds, particularly the primitive ornithuromorphs *Archaeorhynchus spathula* (*left*) and *Xinghaiornis lini* (*below*), display features that straddle the divide between typical enantiornithines and ornithuromorphs, the two main groups of Mesozoic birds. While having the stout feet of a ground-forager common to many other basal ornithuromorphs, the skeleton of *Archaeorhynchus* shows features that are typical of enantiornithines and absent in most other ornithuromorphs (the avian group that includes all present-day birds). Similarly, *Xinghaiornis* exhibits features that are most commonly observed among enantiornithines, even if its long, toothless bill and elongated legs suggest a different lifestyle than the arboreal habits of most enantiornithines.

The degree of intermediacy between the two main groups of Mesozoic birds shown by both *Archaeorhynchus* and *Xinghaiornis* highlights the proximity of these Jehol birds to the evolutionary split between ornithuromorphs and enantiornithines. Additionally, the anatomical differences between these two primitive ornithuromorphs underscore the wide range of lifestyles that evolved during the early history of the group that contains all living birds. The elongate probing bill and relatively long legs of *Xinghaiornis* contrast with the massive beak and shorter legs and toes of *Archaeorhynchus*. These features indicate that *Xinghaiornis* probably foraged along the shores of the ancient Jehol lakes while *Archaeorhynchus* had a more terrestrial, land-dwelling existence.

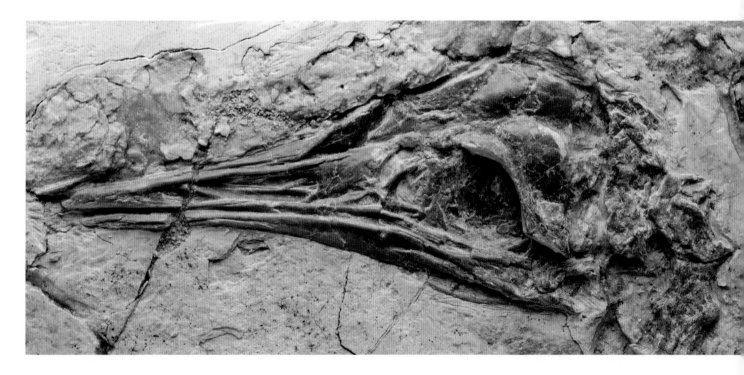

Archaeorhynchus spathula
IVPP-V17075
Jiufotang Formation
Jianchang County
Liaoning Province

Xinghaiornis lini
XHPM-1121
Yixian Formation
Sihetun, Yixian County
Liaoning Province

The hongshanornithids represent one highly specialized group of primitive ornithuromorphs known from a handful of species spanning the entire duration of the Jehol Biota. Unlike other primitive ornithuromorphs, species of this 10-million-year lineage were of small size, roughly comparable to that of a snowy plover. Their delicate legs were long and so were their toes (*left* and *below*), indicating that these animals were well adapted for walking on the muddy shores of the Jehol lakes. Their pointy snouts were equipped with tiny teeth. Clusters of stones presumably contained within a muscular gizzard tell us that the diet of these birds was mainly seeds and other hard food items. The inferred wading habits of these animals also suggest that they might have enjoyed snapping up insects and small invertebrates that lived on the lakeshore.

Some of the fossils of these birds preserve sufficient plumage to enable us to accurately reconstruct the outline of both their wings and feathered tails. These fossils indicate that *Hongshanornis longicresta* (*left*), the group's best-known species, had relatively long and narrow wings, and a long feathered tail capable of fanning out. The proportions of the wings, tail, and overall skeleton of this bird are remarkably similar to those of many small living birds. This correspondence suggests that the ancient hongshanornithids were proficient fliers with flight modes akin to those of some present-day small birds.

Hongshanornis longicresta
IVPP-V14533B
Yixian Formation
Shifo, Ningcheng County
Inner Mongolia

Longicrusavis houi
PKUP-V1069
Yixian Formation
Dawangzhangzi,
Lingyuan County
Liaoning Province

Ph001043

Another group of specialized primitive ornithuromorphs from the Jehol Biota includes the ubiquitous *Yanornis martini* (*left*) and the far less common *Piscivoravis lii* and *Yixianornis grabaui* (*below*). Many fossils of *Yanornis* have given us a good picture of what this bird looked like. With a size similar to that of a Heermann's gull, *Yanornis martini* had a relatively long snout with numerous sharp teeth distributed along most of its powerful jaws (*overleaf*). Its legs were strong and its feet might have been partially webbed, a trait that ornithologists call semipalmate, which helps plovers, sandpipers, and some other shorebirds walk on soft coastlines. Well-preserved plumage in some fossils also tells us that *Yanornis* had powerful wings that together with a long feathered tail were capable of generating significant aerodynamic force. These fossils also show that short feathers covered two-thirds of its shank and that large scales protected its feet. Many fossils preserving the remains of fish inside the digestive tract indicate that *Yanornis* was a superb angler that took advantage of the diversity of fishes living in the Jehol lakes.

Food contents in the stomach of the somewhat larger *Piscivoravis* also show it to be a fish-eater. The slightly smaller *Yixianornis* is known by just one specimen (*below*). While in many ways similar to *Yanornis* and *Piscivoravis*, *Yixianornis* had longer toes and fewer teeth. Nonetheless, the fact that it shares a number of features in common with the other two birds suggests that this less known fossil bird most likely had a similar lifestyle.

Yanornis martini
BMNHC-PH1043
Jiufotang Formation
Lamadong, Jianchang County
Liaoning Province

Yixianornis grabaui
IVPP-V12631
Jiufotang Formation
Qianyang, Yixian County
Liaoning Province

Yanornis martini
BMNHC-PH928
Jiufotang Formation
Linglongta, Jianchang County
Liaoning Province

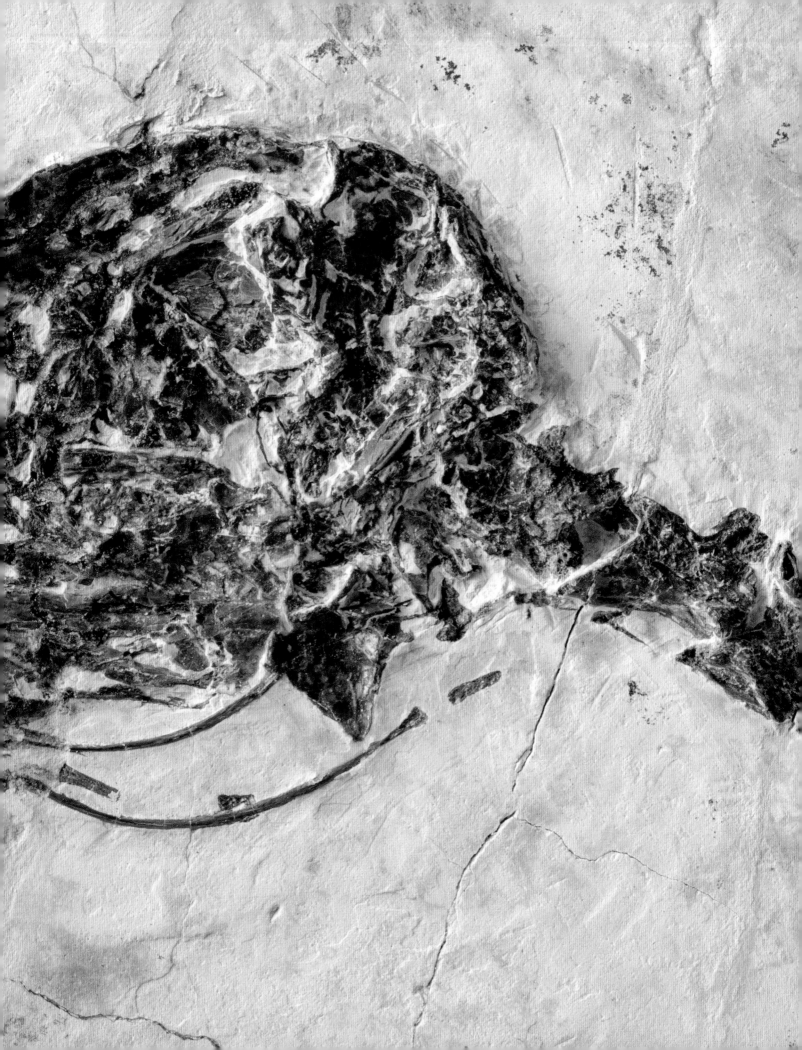

Ph001318

The diversity of birds recorded in the Jehol Biota spans a large portion of the avian family tree and it includes examples that are anatomically similar to present-day birds. While still possessing small and densely packed teeth in the lower jaws and rear portion of the upper jaws (*below*), the 120-million-year-old *Gansus zheni* (*left* and *overleaf*) is the most advanced bird known today from the Jehol Biota. Represented by dozens of specimens, the skeletal architecture of this bird indicates that it was most likely amphibious, foraging at and living in the interface between the shore and the water of the Jehol lakes. Dozens of specimens from a species closely related and contemporaneous to this bird, *Gansus yumenensis*, have been discovered in Early Cretaceous lake sediments from sites representing environmental conditions similar to that of the Jehol but 2,000 kilometers (about 1,250 miles) to the west, in Gansu Province in northern China. These discoveries show that 120 million years ago, gansuids were broadly distributed in what is today northern China and inhabited similar coastal environments of freshwater lakes.

These fossils also help alleviate a longstanding controversy. Using clock-like approaches to calibrate family trees of present-day birds based on the characteristics of their DNA, molecular researchers have consistently claimed that the first modern birds evolved deep in the Cretaceous Period, more than 100 million years ago. The paleontological evidence, however, has failed to demonstrate such an old origin. Fossils identified as belonging to Cretaceous lineages of modern birds are all limited to approximately the last 15 million years of the Cretaceous, between 80 and 65 million years ago. The existence of much older birds that are in many ways similar to their living relatives helps ease this decades-old debate. The anatomical proximity between the 120-million-year-old gansuids and their present-day counterparts thus makes the molecular claim for a deep Cretaceous evolutionary divergence of modern birds more plausible.

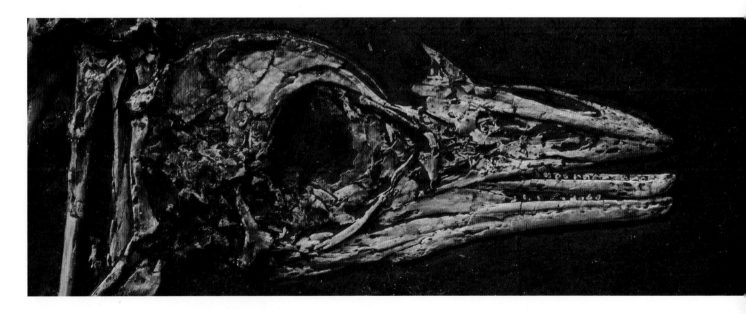

Gansus zheni
BMNHC-PH1318
Jiufotang Formation
Sihedang, Lingyuan County
Liaoning Province

Gansus zheni
BMNHC-PH1392
Jiufotang Formation
Sihedang, Lingyuan County
Liaoning Province

Gansus zheni
BMNHC-PH1343
Jiufotang Formation
Sihedang, Lingyuan County
Liaoning Province

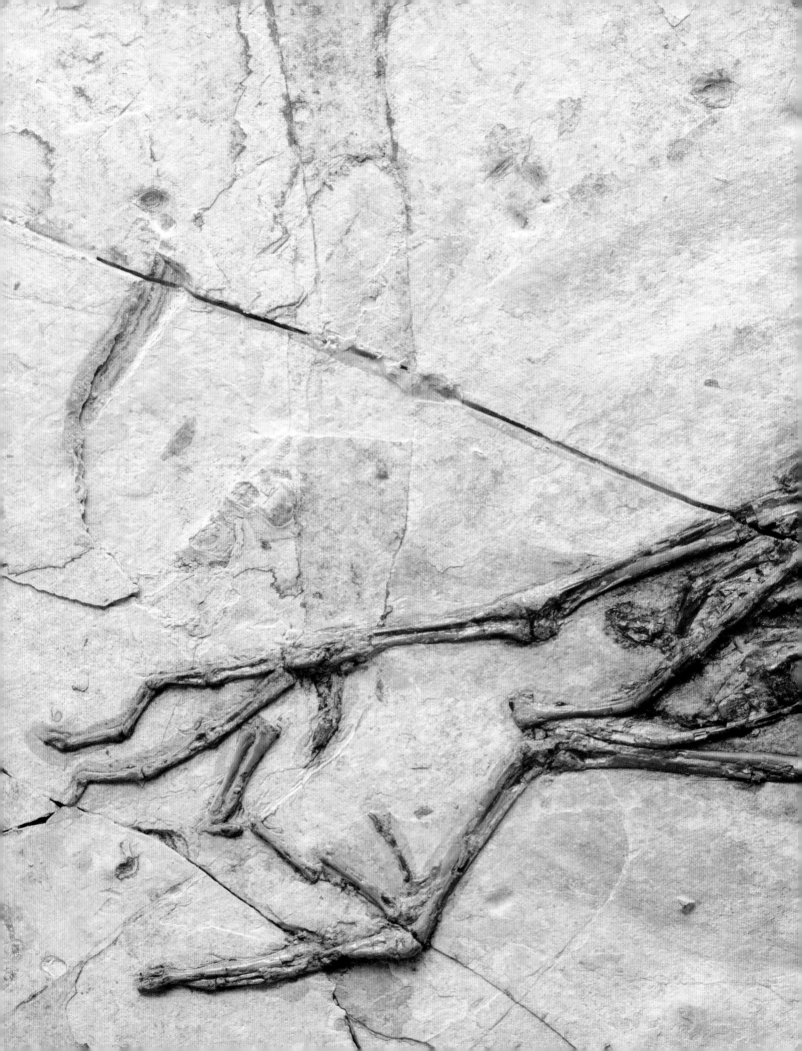

Birds have evolved a unique reproductive system over millions of years. While their embryos develop two ovaries and oviducts, in most birds only the left ovary and oviduct become functional after hatching. A handful of exceptional discoveries of Jehol fossils has opened up a tantalizing window into the reproductive physiology of early birds. These rare fossils contain a cluster of grape-like structures of different sizes clearly inside the body cavity that, while difficult to interpret, have been regarded as nearly mature egg precursors. Moreover, because these structures are interpreted as being preserved only on the left side of the body, they have been used to argue that the degeneration of the right ovary and oviduct of modern birds arose early in the evolutionary history of these animals.

In present-day birds, the quick formation of the yolk inside the eggs leads to a size grading among the developing eggs in which the largest egg is the first to ovulate. The Jehol enantiornithine shown here (*below* and *right*) preserves 14 different-sized, grape-like structures that mostly cluster in front of the bird's hip, and apparently, on the left side. Other Jehol enantiornithines contain fewer and relatively larger egg-like structures. Such evidence, if accurately interpreted, suggests that enantiornithines, and presumably other early birds, had a range of reproductive strategies in which some laid larger clutches of smaller eggs while others had smaller clutches of larger eggs. The common reduction of the right ovary in modern birds has been traditionally interpreted as an adaptation for weight loss, a modification critical for developing enhanced flying capabilities. If correctly interpreted, the discovery of these structures tells us that early in their evolution birds lost the function of one of their oviducts, perhaps as a means of reducing the weight of gravid females.

Enantiornithine indeterminate
STM-29-8
Yixian or Jiufotang Formation
Toudaoyingzi,
Jianchang County
Liaoning Province

Egg size determines many features of the lives of birds, including the wide range of parental dependence exhibited by their chicks. Songbirds and parrots hatch without feathers, their eyes are closed, and they are incapable of leaving the nest. These altricial chicks cannot survive without the care of their parents. In contrast, the precocial hatchlings of ostriches, ducks, quails, and their relatives are quite independent from the moment they are born. They hatch with plumage and are fully capable of finding food. Many present-day chicks have a level of parental reliance in between these two extremes. Determining the degree of parental dependence is difficult for extinct birds because most available fossils represent adults. Fortunately, egg size gives us a hint about the faculties of their hatchlings because among living birds altricial species generally lay eggs that are relatively larger (for a given body size) than those of precocial species.

While few eggs of Mesozoic birds have been discovered, the egg size of these ancient birds can be estimated from the size of the pelvic canal the egg has to pass through when laid. Thus, the well-preserved skeletons of the Jehol birds let us estimate the degree of parental reliance of their hatchlings. The space delimited by the hipbones of a large *Confuciusornis sanctus* (*right*)—with a size between that of an American crow and that of a raven—could accommodate an egg that was at most 23 millimeters (almost 1 inch) in diameter. Yet living altricial birds the size of this *Confuciusornis* specimen would lay eggs that are about 30% larger. Thus, when compared to a wide range of living birds, the size of the pelvic canal of *Confuciusornis* suggests that the chicks of this ancient bird were quite independent and capable of finding their own food—likely also the case for other early birds.

Confuciusornis sanctus
DNHM-D2454
Yixian Formation
Sihetun, Yixian County
Liaoning Province

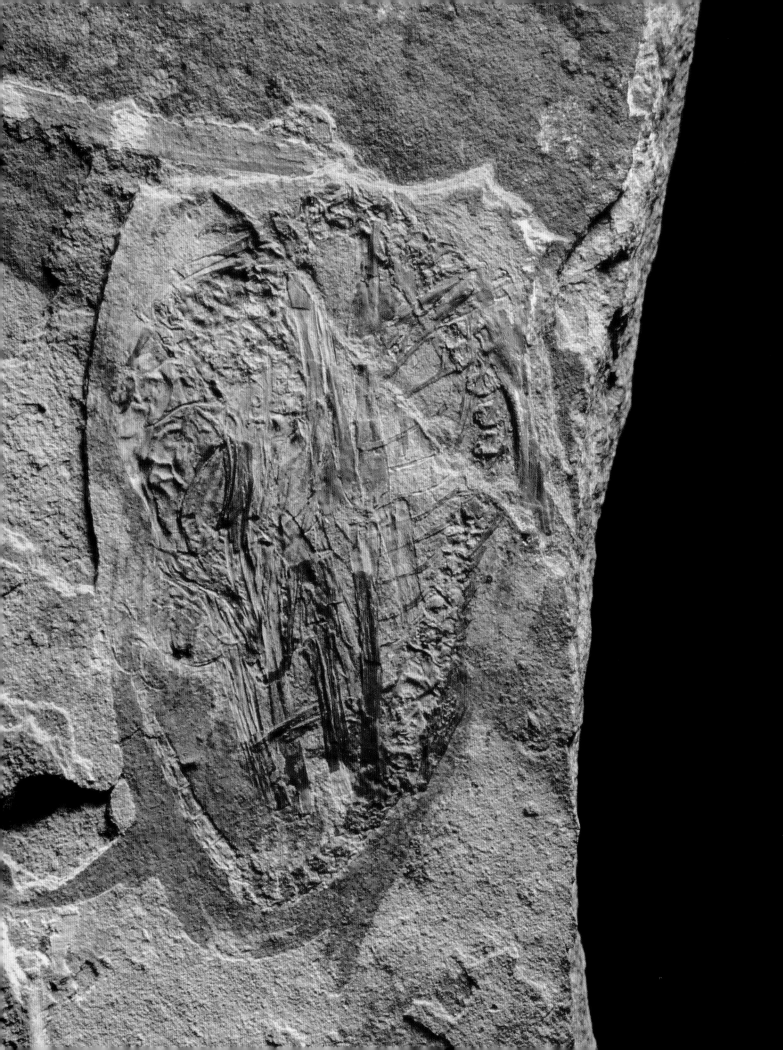

Although fossilized eggs containing the unhatched remains of early birds have been discovered at Cretaceous sites around the world, these fossils remain exceedingly rare. This scant evidence suggests that some Cretaceous birds nested on the ground, close to small streams and other bodies of water. In some cases, the abundance of these eggs led researchers to believe that some early birds were colonial nesters, a pattern commonly seen among living birds, particularly aquatic ones. Surprisingly, the fossil record of bird eggs and embryos from the Jehol is almost nonexistent. No definitive evidence of a shelled egg has ever been found, and only one embryo, belonging to an enantiornithine bird (*left*), has been reported.

Curled inside what appears to be the edge of a 35-millimeter-long (1.4 inches) egg, the body of this tiny embryo is virtually intact, with feathers still enclosed in their growing sheaths. Its well-formed skeleton, with the head between the feet and spine, displays the tucked posture typical of near-hatching embryos of modern birds (*below*). The degree of development of its bones and the presence of well-formed feathers tell us that enantiornithine chicks were capable of moving around in search of food soon after they hatched. Evidence from the estimated size of their eggs and the degree of formation of the bones of hatchlings indicate that such precocial behavior is likely to have been typical for the chicks of other early birds.

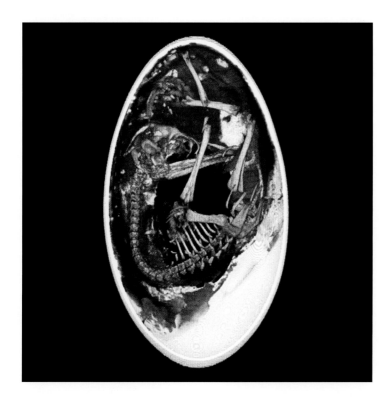

Enantiornithine indeterminate
IVPP-V14238
Jiufotang or Yixian Formation
Liaoning Province

Birds, like most other animals, endure a dramatic transformation from the time they hatch until they become adults. The straight beak and short, chubby legs of a hatchling flamingo are a far cry from the curved beaks and long legs of its elegant parents. Understanding the anatomical transformations that take place as embryos and hatchlings become adults is critical for understanding a wealth of aspects of the lives of birds. Fortunately, the Jehol Biota has given us examples of both an embryo and several hatchlings, even if these are limited to enantiornithine birds. These fossils are recognized for the relative lack of bone formation, which is pervasive in both embryos and hatchlings of living birds, among other specific features such as proportionally large skulls, enormous eye orbits, and short snouts that are typical of the early stages of growth.

The transparent body of a 13-day-old chicken embryo (*below*) with its bones stained in red and the cartilage in blue, shows the large head, enormous orbits, short snout, and reduced wings typical of embryos and hatchlings. These features are reminiscent of those of enantiornithine hatchlings from the Jehol Biota (*right*). The discovery of Jehol enantiornithines representing perinatal stages of development (before and after hatching) has enabled us to better understand the changes in the skeleton of these birds as they grew to adulthood and to compare them to the developmental phases known for living birds. The well-formed skeletons of these fossil hatchlings suggest that enantiornithines might have acquired flight capabilities quickly after hatching.

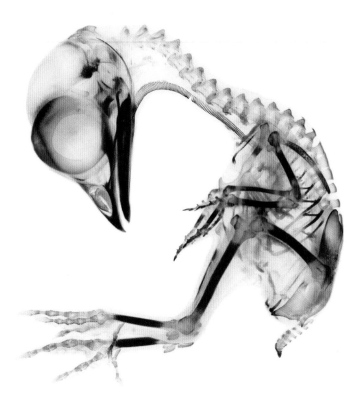

Enantiornithine indeterminate
NIGPAS-130723
Yixian Formation
Dawangzhangzi, Lingyuan County
Liaoning Province

The fact that among the Jehol fossils all known bird embryos, hatchlings, and fledglings are limited to enantiornithines suggests that such a bias may well be related to the specifics of the nesting habits of these and other birds. Why we only have evidence of the early phases of growth for enantiornithines and none for the many other groups of birds represented in the Jehol Biota, including the abundant *Confuciusornis sanctus*, is not known. Perhaps some species of enantiornithines nested close to the lake, thus increasing the possibility that their brood would die and become buried in the lake floor. Nevertheless, the available evidence gives us important information about the growth patterns of primitive birds as a whole.

Young juveniles are often recognized from the proportions of certain parts of the skeleton: they have proportionally large heads with enormous eye orbits and relatively short wings. Furthermore, close examination of their skeletons reveals a large density of pits and groves scarring the surface of their bones (*below*), indicating that the bones were not fully formed at the time of death. Juveniles (*right*) allow us to understand aspects of the development of the skeletons of early birds, how the compound bones that characterize the avian skeleton formed, and when they formed in relation to one another. They also provide critical information for deciphering the development of important parts of the skeleton such as the breastbone, showing the different phases in which this large bone grew over time. Evidence from the plumage of the Jehol hatchlings indicates that early birds were mainly precocial, with hatchlings that might have been able to fly within days, if not the same day, after they hatched.

Enantiornithine indeterminate
GMV-2158
Yixian Formation
Jianshangou, Yixian County
Liaoning Province

A great deal of information about the growth patterns, and overall biology, of extinct animals has come from studying the characteristics of their bone tissues, which are often strikingly preserved in their fossils. Birds are no exception. The stunning fossils of the Jehol birds preserve the characteristics of their bone tissue in exquisite detail. Paper-thin sections of their fossilized bones reveal a wealth of attributes of the microstructure of the bony tissue, thus enabling us to answer a variety of questions related to how fast these birds grew, whether they increased their size continuously, how old they were when they began breeding, and when they reached full-grown size.

A thin section (*below* and *right*) from the radius (a bone of the forearm) of a *Confuciusornis sanctus* shows the intricate nature of this bird's bone microstructure. The characteristics of the bone tissue of this specimen indicate that it was still growing rapidly when it died. These features also tell us that the bird died during its first year of life. Under unusual circumstances, features of the bone tissue can also tell us the sex of a particular individual, even if it died millions of years ago. A type of tissue called medullary bone starts forming inside the central cavity of the long bones of female birds soon before the breeding season. Very rarely, this transient bone tissue is actually preserved in fossils, thus allowing us to identify females that died during their reproductive season. This tissue, however, is not preserved in the fossil shown here and we are thus left to wonder whether this relatively young *Confuciusornis sanctus* was male or female.

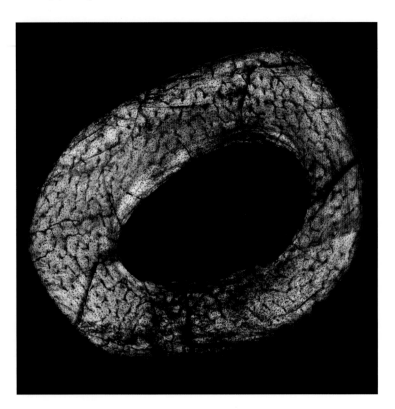

Confuciusornis sanctus
CDL-B-05
Yixian Formation
Sihetun, Yixian County
Liaoning Province

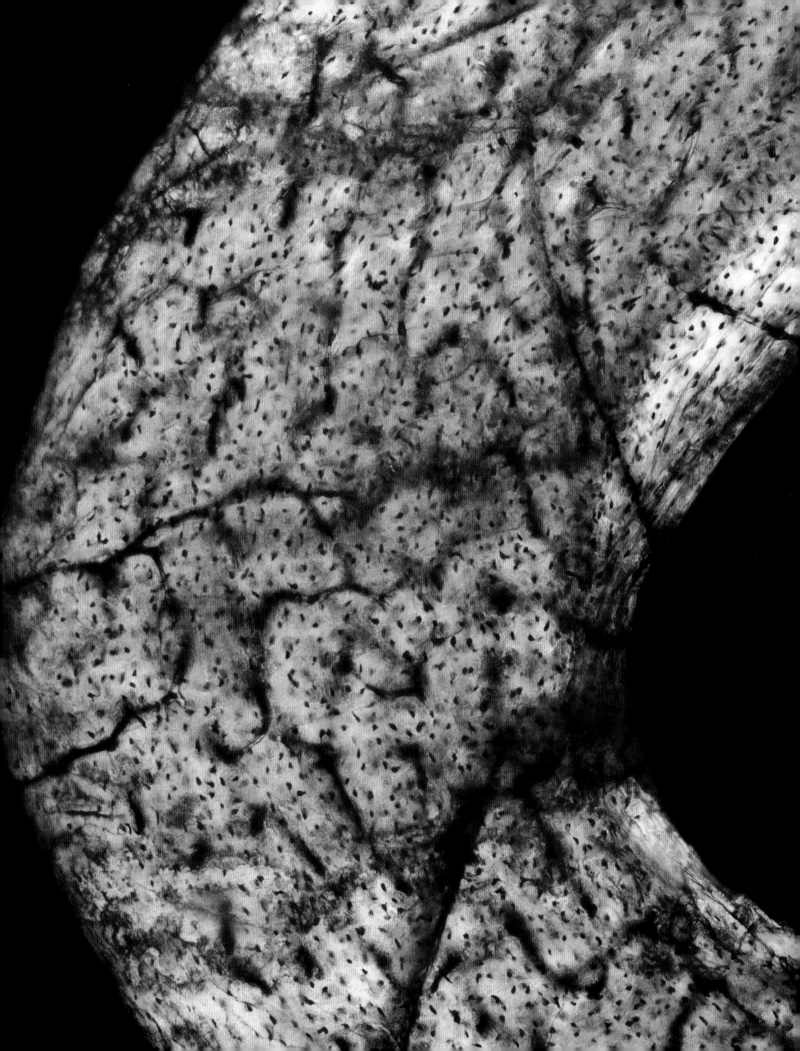

Living birds grow very fast, typically reaching full-grown sizes within a year. Everyone who has ever reared a lovebird has experienced firsthand how the essentially blind and featherless hatchling becomes a raucous adult in just a few weeks. However, our greater understanding of the characteristics of the fossilized bone tissue and the overall developmental pattern of early birds indicate that these animals grew in a very different fashion from their present-day relatives. The Jehol rocks have produced many specimens of the same species of birds—*Confuciusornis sanctus* being the most abundant. These discoveries have shown that *Confuciusornis* and other early birds grew over a prolonged period of time after they hatched. These animals might have quickly developed adult plumage and the physical appearance that differentiated them from hatchlings and fledglings, but they continued to grow in size, scaling up their proportions, for several years.

The two specimens of *Confuciusornis sanctus* shown here (*right* and *below*) are virtually indistinguishable anatomically. Yet the largest (*right*), with a wingspan of about 90 centimeters (3 feet), is 50% larger than the smallest one (*below*). Such a size difference among specimens of a single species with an adult appearance reveals that the normal pattern of growth of these ancient birds was very primitive, and much more like that of today's reptiles such as crocodiles. In this respect, *Confuciusornis* and other very primitive birds had growth characteristics that were similar to those of their dinosaurian forerunners.

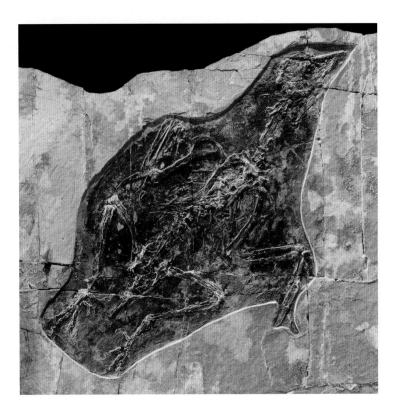

Confuciusornis sanctus
LPM-0233
Yixian Formation
Sihetun, Yixian County
Liaoning Province

Confuciusornis sanctus
DNHM-D2151
Yixian Formation
Sihetun, Yixian County
Liaoning Province

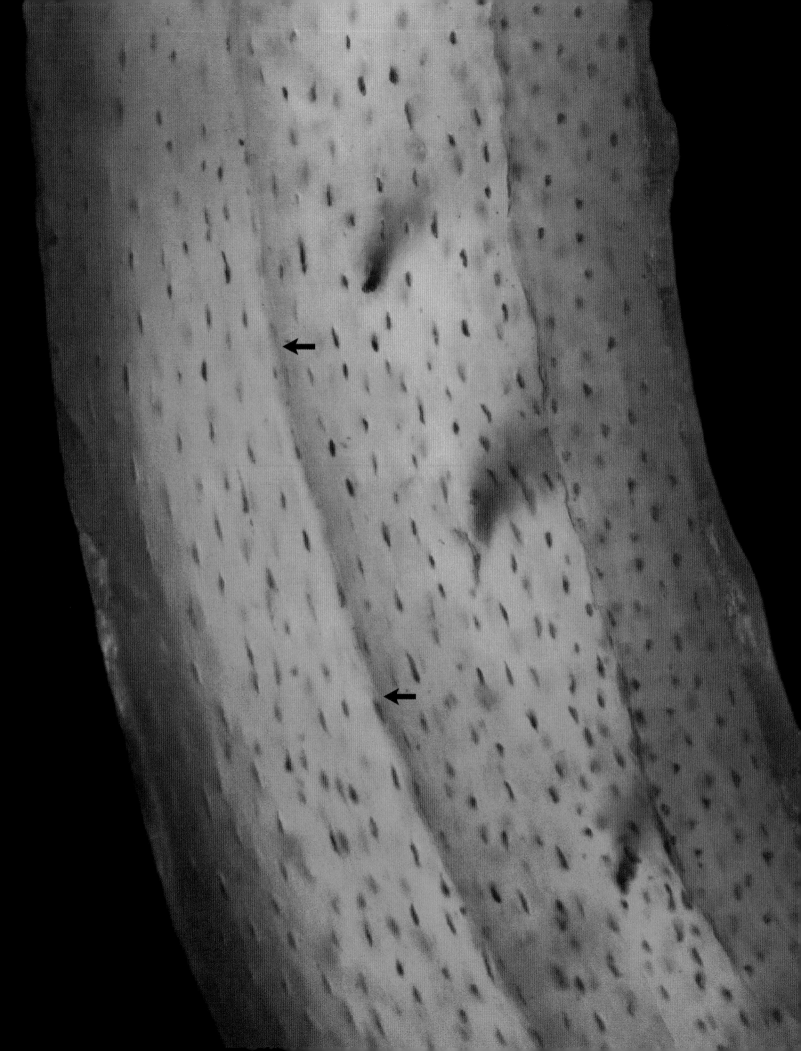

The microscopic features of bone tissue—its density, cellular composition, and so on—are usually well preserved in fossil skeletons. Paper-thin sections of the bones of early birds often show fine concentric lines crossing the main portion of the bone tissue. These lines are interpreted as growth lines, demonstrating pauses in bone growth. Comparisons with living animals whose bones reveal similar growth lines—crocodiles, turtles, and frogs, among others—indicate that these lines are formed cyclically, every year. Such growth rings, however, are rare among present-day birds, which grow very rapidly and almost universally reach adult body size within the first year of life. Yet, the bones of some early birds (including fossils from the Jehol Biota) are known to have several growth rings, indicating that, unlike most of their living counterparts, it took several years until they became full-grown.

A section taken from the shaft of the thighbone—the femur—of the medium-sized enantiornithine *Zhouornis hani* (*below*) reveals the presence of a growth ring (*left, arrows*). Such a growth ring indicates that this individual died in its second year of life. The bone tissue's overall appearance tells us that this animal was still growing, albeit slowly, when it perished. A wealth of detailed studies of the bone tissue characteristics of early birds has allowed us to estimate the ages at which many of these birds died, even if we cannot determine their lifespans. These investigations have produced essential information for understanding the growth rates, age of maturation, and other developmental aspects of these primitive birds.

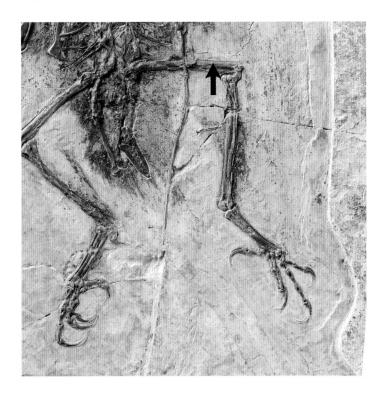

Zhouornis hani
CNUVB-0903
Jiufotang Formation
Chaoyang County
Liaoning Province

Among the most eye-catching features of *Confuciusornis sanctus* is a pair of extremely long feathers jutting out from the tail of this beaked bird. Hundreds of well-preserved fossils have shown that not all specimens carried these feathers (*right*). Fossils found in exactly the same stratigraphic level, and even on the same slab, indicate that individuals with and without these feathers lived next to one another. Close examination of fossils lacking these tape-like feathers reveals that their absence is not due to preservation; in many instances, the plumage of these fossils is pristinely preserved with no evidence to suggest that these stiff feathers could have been lost during the fossilization process. Clearly, the fact that some fossils display these long tail feathers while others do not tells us that these feathers were ornamental—a sexual trait that distinguished males from females engaged in courtship.

In addition, the discovery of a type of bone tissue unique to egg-laying hens of present-day birds in a fossil of *Confuciusornis sanctus* without ornamental feathers has validated this interpretation: at least a fossil without ornamental feathers was shown to be that of a female. Called medullary bone, the type of bone tissue found in this unique fossil is a singular adaptation of female birds that provides a reservoir of calcium to be recycled at the time an egg is enclosed in its protective eggshell.

The discovery of medullary bone in a specimen of *Confuciusornis sanctus* lacking ornamental feathers was also instrumental for understanding the onset of puberty in early birds. Among other things, the discovery documented that, unlike their living relatives, the primitive birds of the Mesozoic Era matured sexually well before reaching their full-grown size. In fact, some of the specimens of *Confuciusornis sanctus* might have become reproductively active when they weighed only one-fourth of the weight reached by the largest known specimens, a breeding strategy unknown for any modern bird. This example shows how the sheer number of fossils of Jehol birds and their extraordinary preservation have provided information for addressing a diversity of evolutionary and ecological questions about the origin of the unique attributes of living birds.

Confuciusornis sanctus
NIGPAS-139379
Yixian Formation
Sihetun, Yixian County
Liaoning Province

From the majestic tail of the peacock to the fancy plumes of the birds-of-paradise, feathers constitute prime examples of features involved in courtship and sexual display. A large number of Jehol birds, including *Confuciusornis sanctus* and its kin as well as many enantiornithines, are known to have carried a pair of long and stiff feathers projecting from the end of their bony tails (*below*). The slender profile of these feathers could have hardly generated any aerodynamic forces (*left*). Instead, ideally designed to impress, these feathers have been consistently regarded as ornaments.

Reaching in some cases more than 40 centimeters (16 inches) in length, as in some large specimens of *Confuciusornis*, these long feathers regularly exceed the length of the bodies of the birds that carried them, and in some cases they are several times as long (*overleaf*). Subtle differences in their appearance and proportions, either in their length with respect to the body or in the characteristics of their vanes and shafts, suggest that these ornamental feathers were specific for each of these bird species. The diversity of ornamental plumages documented by the Jehol birds tells us of a wealth of sexual displays evolving early in the evolutionary history of birds. Among living birds, there are many examples of flamboyant plumages involved in stunning displays that never cease to amaze us. One can hardly imagine the choreography of courting behaviors of the avian repertoire from the Jehol that are sadly not recorded in the fossil record.

Enantiornithine indeterminate
BMNHC-PH1154A
Huajiying Formation
Sichakou, Fengning County
Hebei Province

Enantiornithine indeterminate
BMNHC-PH807
Yixian Formation
Jianchang County
Liaoning Province

Junornis houi
BMNHC-PH919A
Yixian or Jiufotang Formation
Liutiaogou, Ningcheng County
Inner Mongolia

Coloration plays a fundamental role in the lives of birds. It helps them camouflage themselves and avoid being preyed on, provides the basis for a diversity of display behaviors, and is paramount for their reproductive strategies. Paleontologists have long wondered about the colors that adorned the bodies of dinosaurs and the early birds. Yet, identifying the color of extinct animals has been one of the most difficult problems in paleontology and only recently have approaches to this end been developed.

One approach that has made significant headway in deciphering the color of the plumage of extinct birds has looked at the distribution of melanosomes—tiny capsules that contain the different types of the dark brown to black melanin pigments (*below*)—as a proxy for the distribution of melanin-based colors in feathers. This approach assumes that the greater the concentration of melanosomes found in a fossilized feather, the darker the color of the feather. Additionally, the shape of the melanosomes gives us a hint of the specific tonalities that colored the ancient feather because their particular shape is related to the specific types of melanin they carry.

Another sophisticated approach has used traces of copper and other metals as proxies for coloration. In this case, fossils are studied through synchrotron technology, using a particle accelerator to reconstruct a color map of the plumage based on the types of trace metals that are used as proxies for the density of melanin in fossilized feathers. While much remains to be fine-tuned about these methods, neither of them uncontroversial, these novel approaches have the potential of giving us glimpses of the colorful plumages that adorned the bodies of the ancient Jehol birds.

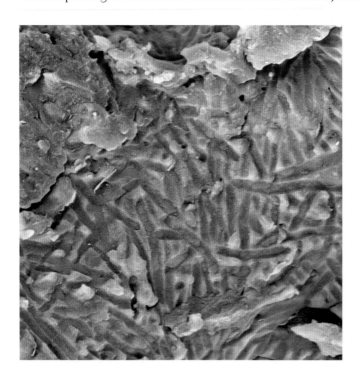

Enantiornithine indeterminate
BMNHC-PH1061B
Huajiying Formation
Sichakou, Fengning County
Hebei Province

Feathers are the most elaborate skin outgrowths of any vertebrate animal. Numbering many thousands in a single bird, these remarkable structures serve a variety of functions, from insulation and waterproofing to aerodynamics and sexual display. Over the long evolutionary history of birds, the light weight, strength, and flexibility of keratin—feathers' building material—have facilitated the evolution of the enormous diversity of sizes and shapes that we see among these structures. Most fossils of Jehol birds preserved portions of plumage, and details of their feathers are often revealed in exquisite fashion (*below* and *right*).

These fossils provide information about a wealth of characteristics, including the length and width of the shafts, the delicate arrangement of the barbs, and whether they are connected to one another by the same system of barbules that characterizes the Velcro-like cohesion of many modern feathers. They also tell us details such as the number of primary feathers attached to the hand bones, the secondary feathers attached to the forearm, and the ones attached to the tail. The superbly preserved plumage of many Jehol birds often helps us estimate the contour of their wings, the shape of their tails, and the outlines of their bodies, features key for assessing flight performance. The configuration of the feathers, and of the airfoils they formed, contribute to our understanding of the different flight modes these birds had millions of years ago.

Enantiornithine indeterminate
BMNHC-PH1061B
Huajiying Formation
Sichakou, Fengning County
Hebei Province

Protopteryx fengningensis
BMNHC-PH1060A
Huajiying Formation
Sichakou, Fengning County
Hebei Province

The exquisite preservation of the Jehol Biota reveals details of archaic feathers that no longer exist today (*left* and *overleaf*). Many of these birds carried a pair of long feathers projecting from the tail. Superficially resembling the racket-shaped feathers that adorn the tail of living motmots, these ancient feathers had a stiff, belt-like shaft that extended for three-quarters or more of the length of the feather. This broad shaft tapered toward the feather's tip, giving way to a vaned portion in which the length of the barbs increased gradually toward the rounded end of the feather (*overleaf*).

Another type of possibly extinct feather known from *Confuciusornis sanctus* and its kin, enantiornithines, and other early birds is a body feather lacking a distinct shaft. In this case (*left*), the feather's barbs differentiate from a central portion lacking a well-defined shaft. These feathers might have played a role insulating the bird's body, like modern down feathers, but they differ from present-day down feathers in that their loosely arranged barbs do not originate from a short shaft.

Recent molecular and developmental studies have shed light on how these extinct feathers might have formed. By experimentally tinkering with the genes that control feather growth in living chickens, we are becoming more aware of the genetic manifestation that millions of years ago regulated the development of feather types that today no longer exist outside the laboratory. These studies show that the ornamental feathers of these archaic birds are highly modified shafted feathers. The fact that such feathers display different relative sizes and appearances in confuciusornithids and enantiornithines indicates that these ornaments were specific to their own species; this specificity most likely reflects unique developmental pathways in feather formation. These laboratory experiments have given us valuable information about the genetic evolution that millions of years ago led to the formation of feathers that disappeared together with the birds that bore them. We can only wonder what other types of feathers covered the bodies of the ancient Jehol birds. Future detailed studies of their plumage, as well as new exceptional fossils, may reveal a greater variety of feathers that disappeared long before the origin of modern birds.

Enantiornithine indeterminate
BMNHC-PH1061B
Huajiying Formation
Sichakou, Fengning County
Hebei Province

Enantiornithine indeterminate
BMNHC-PH869B
Yixian Formation
Ningcheng County
Inner Mongolia

The shape and structure of the wing is the key to the flight competence of birds because these features maximize the relationship between the lift, speed, and drag generated during flight. Modern birds display an astonishing array of wing shapes. Nonetheless, these can be sorted into a few basic types, even if some fall in between these categories. Most songbirds have elliptical wings best suited for maneuverability. Birds that glide and soar have broad and relatively long wings, which confer substantial lift. The narrow and tapered wings of swifts, swallows, terns, and falcons are designed for rapid flight, and seabirds such as albatrosses and gannets have long and slender wings specialized for lift and speed.

Determining with precision the shape of the wings of extinct birds is difficult because even when fossils preserve their plumage in great detail, the wings are often folded or obstructed by other parts of the bird. Nonetheless, the spectacular fossils of the Jehol Biota allow us to make approximations of the shape of the wings of these animals for a number of groups. Well-preserved specimens of *Sapeornis chaoyangensis* (*left*) and *Confuciusornis sanctus* (*overleaf*) let us reconstruct with a certain degree of confidence the outline of the wings of the most primitive short-tailed birds. These fossils tell us that *Sapeornis* had relatively broad wings that reached a span of about 110 centimeters (44 inches) in full-grown individuals. Such wings suggest an ability to glide at relatively low speeds. In *Confuciusornis*, the wings were narrower and proportionally longer, with enormous feathers attached to the hand bones and tapered wingtips. This type of wing suggests a greater capacity for high-speed flight. These generalizations, however, do not take into account the very primitive anatomy of the bones of the shoulder and forelimb of these birds, which makes us wonder whether they actually moved their wings (and flew) in a manner similar to that of living birds.

Sapeornis chaoyangensis
DNHM-D3078
Yixian Formation
Jianchang County
Liaoning Province

Confuciusornis sanctus
HGM-41HIII0401
Yixian Formation
Sihetun, Yixian County
Liaoning Province

THE AVIAN FOSSILS OF JEHOL

Fossils of Jehol enantiornithines show a substantial degree of modernization of the flight apparatus when compared to the bones of the wing and shoulder of more primitive short bony-tailed birds. The proportion of the wing bones of enantiornithine birds (*left*) is similar to that of their living counterparts. Their shoulder bones have been modified to the point of approaching the appearance of those of present-day birds except for a few details. The breastbones are large and they have a shallow keel that expands the origin of the powerful breast muscles that power the wing beat. Additionally, enantiornithines evolved an alula, a small tuft of feathers attached to the thumb that plays a key aerodynamic role, allowing birds to fly at low speeds when needed.

All these innovations with respect to more primitive birds tell us that in terms of flight competence, enantiornithines would have probably been close to indistinguishable from many small birds of today. A few fossils of these birds have well-preserved feathered wings that are elliptical in shape, characteristic of birds that live in densely vegetated environments such as forests. The small to medium size of most enantiornithines coupled with the design of their skeletons and wings (*below*) indicate that these birds were ideally adapted for the maneuverability required for flying in the close environments of the ancient Jehol forests.

Enantiornithine indeterminate
BMNHC-PH1154A
Huajiying Formation
Sichakou, Fengning County
Hebei Province

Enantiornithine indeterminate
BMNHC-PH925
Yixian or Jiufotang Formation
Liaoning Province

The wing of modern birds is a highly modified structure in which the reduced fingers are completely encased in a fleshy sleeve. The skinfolds that encircle the wing bones also provide key aerodynamic functions. In the front, spreading between the shoulder and the wrist, a large triangular skinfold called the propatagium increases the airfoil surface of the wing responsible for generating lift. In the rear, a long and fleshy skinfold running along the forearm and the hand—the postpatagium—anchors the strong shafts of the flight feathers that form the main airfoil of the wing (*below*). Fossils from the Jehol Biota and elsewhere show that birds living as far back as 131 million years ago had evolved essentially the same configuration of skinfolds (*right*).

Microscopic studies of rare fossils have even shown the detailed architecture of the tiny muscles, tendons, and ligaments that fasten to the quills of flight feathers and control their movements. These exceptional fossils tell us that the long fingers of enantiornithines, and presumably those of other early birds, were largely contained within the fleshy tissue of the hand, with only their claws sticking out. Confined to this fleshy casing, these fingers would have been incapable of performing the grasping functions of birds' dinosaurian forerunners. These fossils definitively show that early in the evolution of birds, the anatomical design of the hand underwent fundamental modifications that transformed it into what is essentially the same structure we see today among their living relatives.

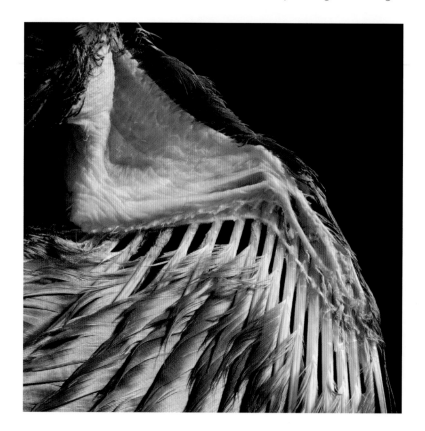

Enantiornithine indeterminate
BMNHC-PH1156A
Huajiying Formation
Sichakou, Fengning County
Hebei Province

Fossils from the Jehol and elsewhere provide abundant evidence that many of the primitive birds of the Mesozoic were reasonably good fliers. These fossils reveal that early in their history, birds evolved a key adaptation that has contributed to their ability to become masters of the air. Although rarely well preserved, several Jehol enantiornithines show the presence of an alula, a structure also known as the bastard wing. Consisting of a tuft of usually four small feathers attached to the tip and length of the thumb (*below*), the alula constitutes the main anti-stalling device that allows birds to fly and maneuver at low speeds. When flying slowly, particularly during landing or when taking off, birds position their wings at a higher angle with respect to the direction of the incoming airflow. This action increases lift production, which is otherwise limited when birds fly at low speeds.

Beyond a certain point, however, the angle between the wing and the incoming airflow generates sufficient turbulence to make the bird stall. This is when the alula comes into play; its deployment creates a slot with the edge of the wing that allows birds to angle their wings even farther and to increase the wing's lift production. The alula's deployment thus enables birds to land safely and to fly at low speeds without stalling. More primitive birds such as *Confuciusornis* and *Sapeornis* seem to have lacked an alula, even though their long thumb might have acted as a rudimentary equivalent. The evolution of this key aerodynamic feature in enantiornithines (*right*) and in primitive ornithuromorphs (the earliest forebears of present-day birds) speaks of a significant refinement in the aerial skills of these ancient birds.

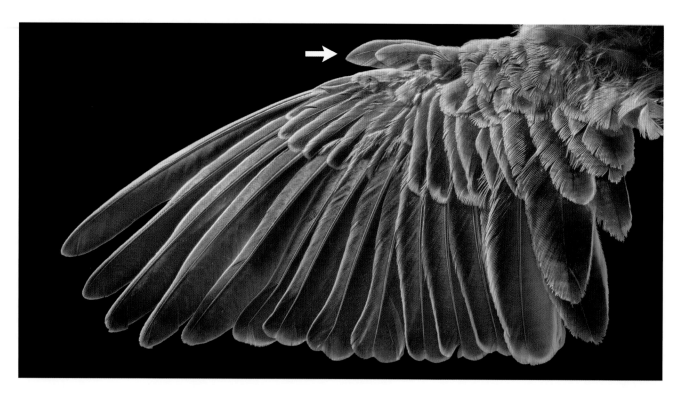

Longipteryx chaoyangensis
BMNHC-PH1071
Jiufotang Formation
Lamadong, Jianchang County
Liaoning Province

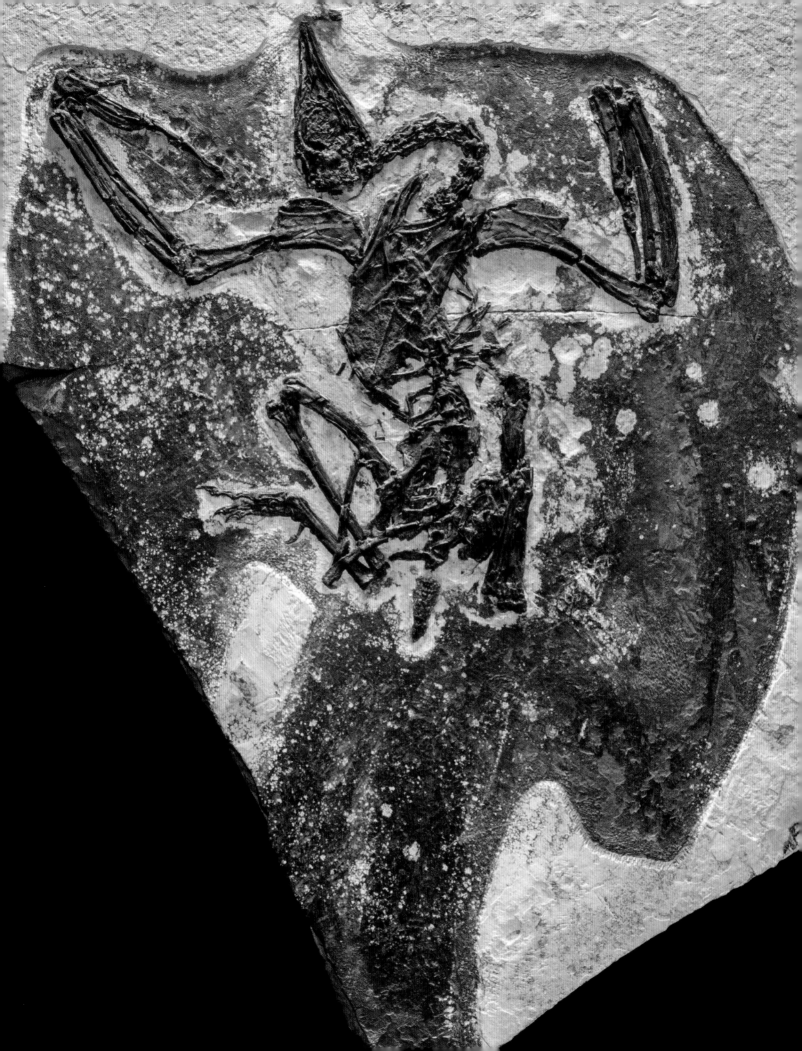

The shape and the size of the feathered tail of birds vary considerably among groups. In most species, the tail feathers are primarily responsible for braking and steering but they can also generate additional lift. Some birds, however, have greatly modified tails whose main function is to show off. The magnificent avifauna of the Jehol Biota gives us a hint of the diversity of feathered tails that characterized the birds that lived during the first half of the Cretaceous Period. The ornamental plumage of *Confuciusornis sanctus* and its kin as well as the long tail feathers that jut off the rump of many enantiornithines had no role in the flight abilities of these birds. Instead, they were largely designed for sexual display. Fossils of Jehol ornithuromorphs, however, show that several types of aerodynamic tails evolved among these early relatives of modern-day birds.

Albeit poorly preserved, the tail of *Schizooura lii* (*left*) appears forked, its outer feathers separated by a deep central notch, resembling the tail of some living birds, including various flycatchers, swallows, and terns. Better-preserved fossils of *Hongshanornis longicresta* (*overleaf*) indicate that this bird had relatively long and vaned feathers forming a rounded tail capable of some fanning, and thus playing a more significant role in lift generation. Fossils of *Yanornis martini* (*below*) show that this relatively common Jehol bird had a tail that could fan out into a wider surface, thus generating more significant lift forces. The diversity of feathered tails of these early ornithuromorphs, while modest, tells us that tail designs typical of numerous modern birds evolved early in the history of the group.

Schizooura lii	*Yanornis martini*	*Hongshanornis longicresta*
IVPP-V16861	BMNHC-PH928	DNHM-D2945
Jiufotang Formation	Jiufotang Formation	Yixian Formation
Jianchang County	Linglongta, Jianchang County	Lingyuan County
Liaoning Province	Liaoning Province	Liaoning Province

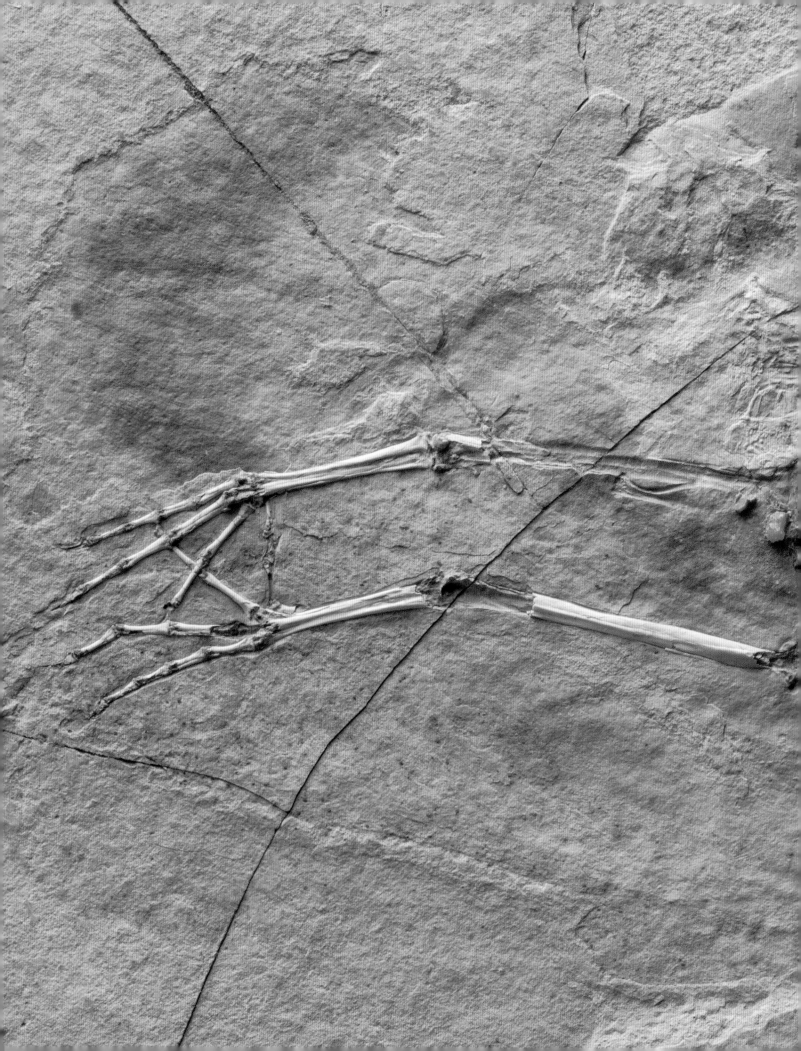

Feathers cover the shank of a great variety of living birds, either partially or completely, and in some groups—grouse and owls, among others—the leg plumage extends to their feet. Leg feathers are thought to have multiple purposes, from aerodynamics in hawks and other predatory birds to insulation in the spruce grouse (*below*) and other birds that live in cold places. Not surprisingly, feathers extending to different levels of the leg have also been documented in some fossil birds of the Jehol Biota (*left*). In general, the leg plumage of these archaic birds consists of simple feathers similar to, but usually shorter than, those covering the rest of their bodies.

Most typically, as in *Confuciusornis sanctus* and a variety of enantiornithines and ornithuromorphs, these feathers cover the upper and middle portions of the shank. In these birds, the lower shank and feet are covered by scales, which are preserved in a handful of exceptional fossils. In some cases, however, the leg plumage extends down to shelter the upper portion of the foot bones (*left*), although there is no known example of a Jehol bird with feathers covering its toes. While we cannot be sure of the role these feathers played in the lives of these ancient birds, the fact that they are short and lacked the cohesiveness of flight feathers indicates that they did not have an aerodynamic function.

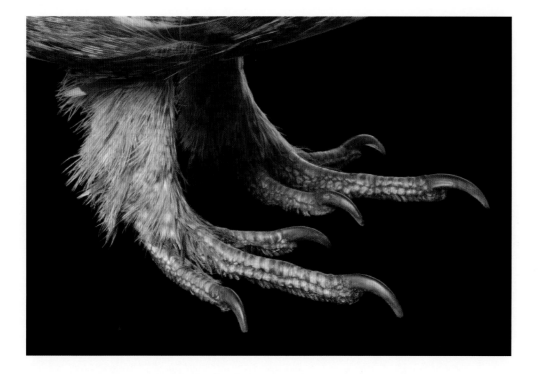

Enantiornithine indeterminate
BMNHC-PH1061B
Huajiying Formation
Sichakou, Fengning County
Hebei Province

All living birds are toothless, but teeth of different shapes and sizes were widespread among their Mesozoic predecessors. These teeth were placed in sockets on all or some of the tooth-bearing jawbones (*left* and *below*). Details of these teeth tell us that they were replaced regularly throughout the animal's life. Deep inside its socket, a new tooth would form and develop. As it grew bigger, the root of the old tooth inside the socket became resorbed to a point when, lacking its anchoring root, the old tooth was shed and the new tooth replaced the old one. Evidence of this process is rare, but certain fossils show a small pit on the interior side of the tooth (*left, arrow*) that points to the area where the root of the old tooth began to resorb. This tiny window may even show the tip of the crown of the newly formed tooth growing inside (*left*). The discovery of such pits in different groups of early birds tells us that these animals had a similar mechanism for replacing their teeth.

Experimental studies on living birds have shown that the genes responsible for the generation of teeth can be artificially reactivated; under special conditions, chickens and other birds develop simple, conical teeth. These studies have also shown that the development of teeth can be activated, or blocked, in some areas of the jaws, or on some specific bones, through tinkering with certain genes. These investigations help us understand the evolution of the different dental configurations present in the jaws of Mesozoic birds. They also show that part of the genetic toolkit responsible for generating teeth in birds has remained dormant for millions of years.

Yanornis martini
BMNHC-PH1043
Jiufotang Formation
Lamadong, Jianchang County
Liaoning Province

Sapeornis chaoyangensis
HGM-41HIII0405
Jiufotang Formation
Dapingfang, Chaoyang County
Liaoning Province

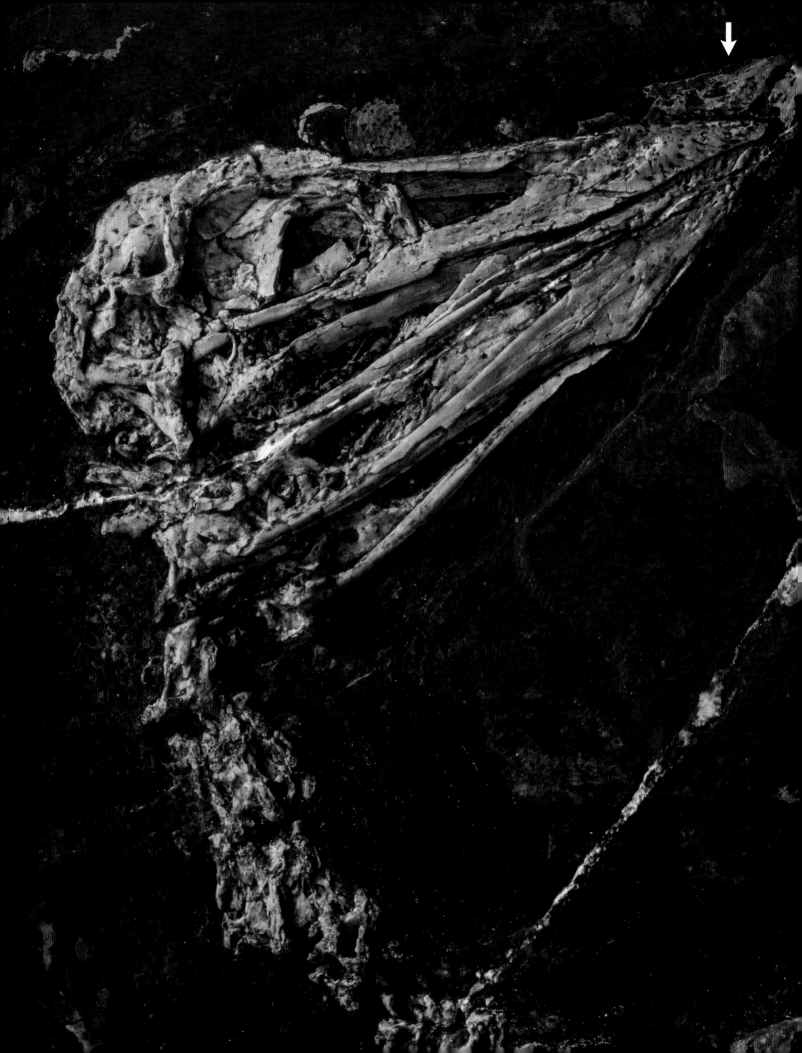

The jawbones of living birds are lined by a durable, horny sheath that forms the bill of these animals (*below*). The astonishing diversity of bill shapes suggests that the evolutionary loss of teeth, widespread among archaic Mesozoic birds, might have triggered the evolution of many different types of beaks and their specific purposes. The fossil record tells us that complete tooth loss was a recurrent phenomenon that took place independently in a number of early lineages of birds. Teeth gave way to beaks in no less than five separate lineages, including *Confuciusornis* and its kin (*left*) but not counting the event that led to tooth loss in the ancestor of modern birds. Why bills evolved multiple times and became so successful is a matter of contention. Traditionally, it has been argued that tooth reduction contributed to weight reduction, a clear advantage for airborne animals. Yet, the weight of the often tiny teeth of many Mesozoic birds appears negligible; thus, tooth reduction would not be a meaningful way to lighten a body.

Most likely, teeth were lost as the birds' digestive system became specialized for food processing. Recent genomic studies show that modern birds lost their teeth more than 115 million years ago, and that such an event evolved in concert with a horny beak. The presence of a horny sheath in fossils is often inferred by a series of closely packed pits scarring the surface of the jawbones, which in living birds allows the passage of nerves and blood vessels that nurse the horny bill. Rare fossils from the Jehol sometimes reveal remnants of this horny structure (*left, arrow*).

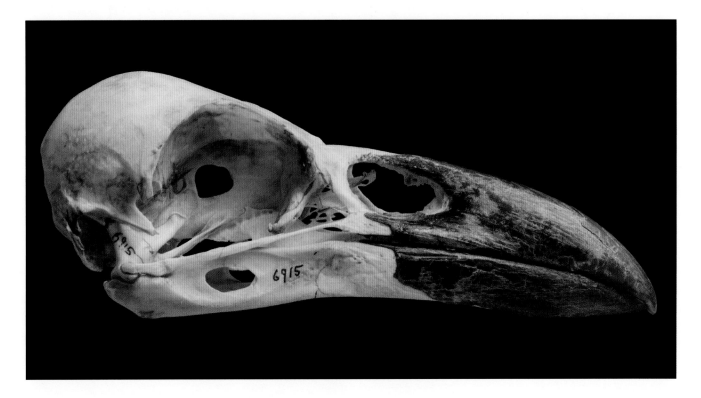

Confuciusornis sanctus
BMNHC-PH986
Yixian Formation
Sihetun, Yixian County
Liaoning Province

Most early birds had teeth, but their structure tells us that they did not use them for chewing. Living birds also do not break down their food in their mouths; they swallow their food whole. Their elaborate digestive tract includes a two-part stomach. In the first part, a glandular organ is used for chemical maceration; the second, muscular part, the familiar gizzard, is dedicated to physically grinding up food. Most enzymatic digestion takes place in the bird's small intestine. To assist the gizzard in breaking food items into smaller pieces and to enhance digestion, many birds regularly swallow grit and other small rocks that reside temporarily inside this organ.

A number of Jehol bird fossils, including ornithuromorphs such as *Gansus zheni* (*left*) and *Archaeorhynchus spathula* (*below* and *overleaf*), have been found containing clusters of angular and poorly sorted stones inside their body cavities, usually in front of their hipbones. These stones are typically located in an area of the body consistent with the position of the gizzard in present-day birds, thus suggesting that these early birds had both a gizzard and the behavior of ingesting grit. Experiments with living birds and mammals have shown that there is no difference in the efficiency of how gizzards and chewing teeth grind food—early birds did not have the powerful dentition of either mammals or many of their dinosaurian forerunners, but they were clearly able to process food with comparable efficiency.

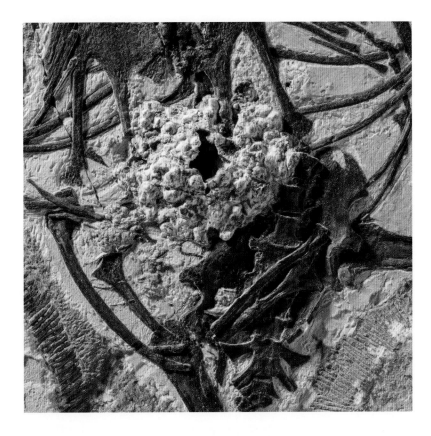

Gansus zheni
BMNHC-PH1342
Jiufotang Formation
Sihedang, Lingyuan County
Liaoning Province

Archaeorhynchus spathula
IVPP-V17075
Jiufotang Formation
Jianchang County
Liaoning Province

Archaeorhynchus spathula
IVPP-V14287A
Yixian Formation
Yixian, Yixian County
Liaoning Province

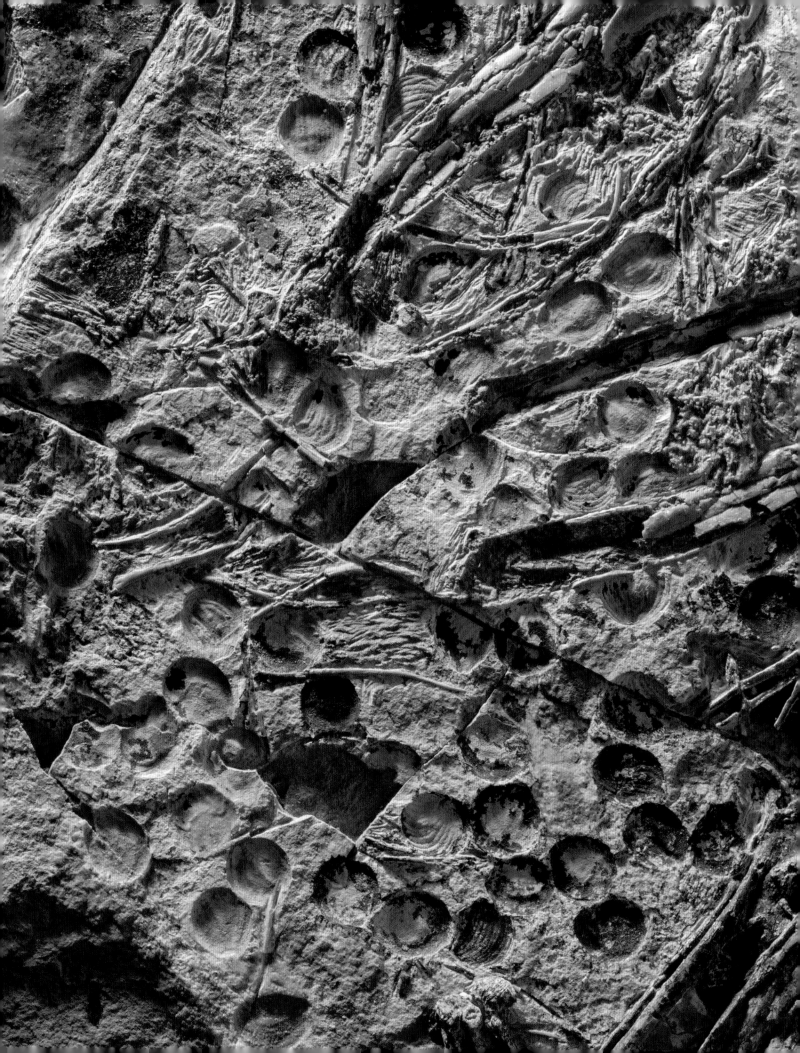

In addition to a two-part stomach, the sophisticated digestive system of birds includes an expansion of the esophagus called the crop (*below*), which allows birds to store food that cannot be digested at the same pace it is consumed. Food in the crop undergoes softening in preparation for its maceration, grinding, and ultimate digestion in the stomach and the small intestine. The existence of a crop has been indirectly documented for a number of Jehol birds. In these fossils, the presence of such a structure can be inferred from a mass of food items that is sometimes preserved beneath the neck, in the location of the esophagus. Taking advantage of rich resources in the ancient Jehol forest, many of these birds ate seeds and fruits (*left*). Remnants of these nutritious foods are sometimes well preserved inside their bodies. In some cases, as in the specimen of *Sapeornis chaoyangensis* shown here (*overleaf, arrows*), a cluster of seeds or fruits is preserved in front of the shoulder bones and wishbone, indicating the presence of a pouch-like crop. The existence of a crop together with evidence indicating the partitioning of the stomach into an anterior glandular organ and a posterior grinding gizzard suggests that early in their evolution, birds acquired an elaborate digestive tract that was probably comparable in architecture, and in efficiency, to that of their present-day relatives.

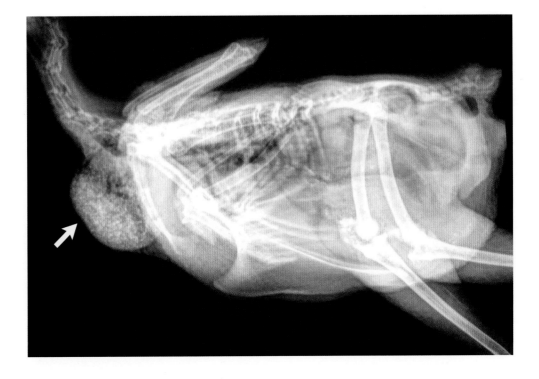

Jeholornis prima
IVPP-V13274
Jiufotang Formation
Dapingfang, Chaoyang County
Liaoning Province

Sapeornis chaoyangensis
BMNHC-PH1067
Yixian or Jiufotang Formation
Linglongta, Jianchang County
Liaoning Province

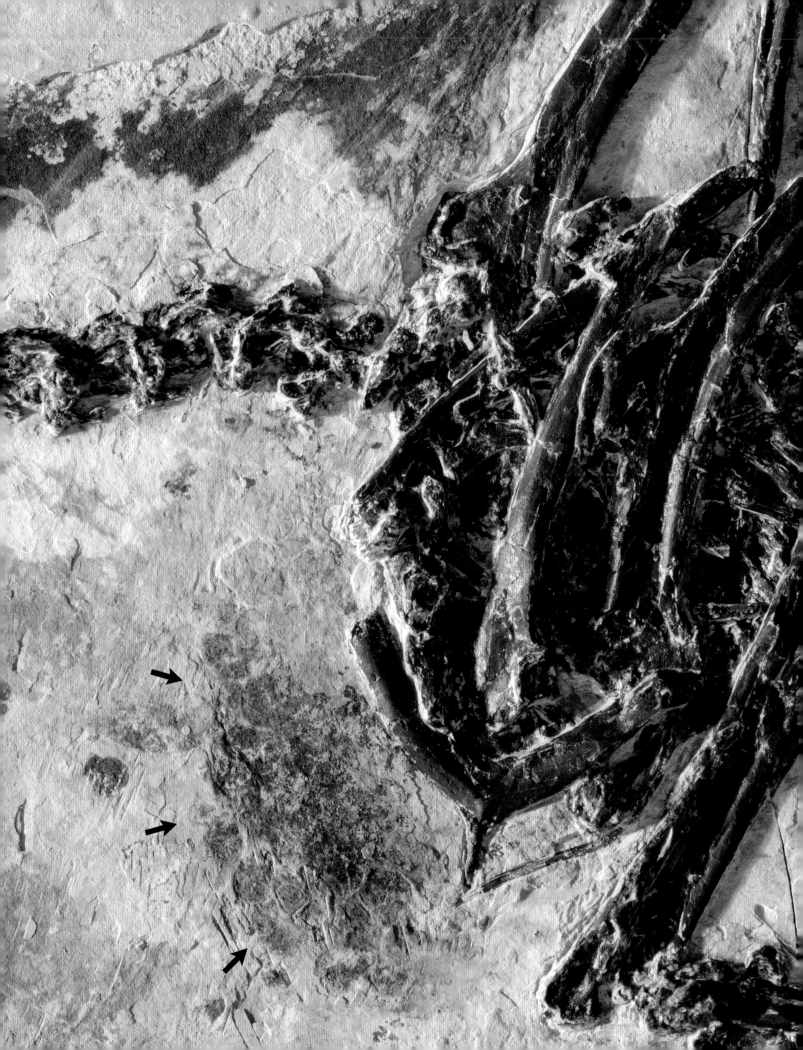

The crop of birds varies in size and shape, and also in its location within the esophagus. Birds that eat grains usually have pouch-like crops located at the base of the neck. Ducks have longer, spindle-shaped crops, and birds that eat fish have crops formed by simple expansions of the entire esophagus. Some fish-eating birds from the Jehol Biota, particularly *Yanornis martini* (*left*), show clear evidence of this type of crop. Among the many fossils of this early ornithuromorph, some preserve numerous disassociated remains of fish (*below* and *overleaf*) along the entire length of the neck and even following the digestive tract beyond the esophagus. Sometimes whole fish have been found preserved lengthwise along the neck of these birds, indicating that, as in many fish-eating birds of today, *Yanornis* had a flexible esophagus capable of storing a sizable catch. The presence of fish-filled digestive tracts in fossils of *Yanornis* tells us that this bird foraged far from its roosting and nesting sites, as many of today's fish-eating birds. It is possible that like most of its living analogues, *Yanornis* carried food in its crop to be regurgitated when feeding its brood. If so, the presence of this behavior also documents that some early ornithuromorphs had adult-dependent nestlings that could not survive without the care of their parents.

Yanornis martini
DNHM-D3069
Yixian or Jiufotang Formation
Liaoning Province

Yanornis martini
XHPM-1205
Yixian Formation
Jinzhou, Yixian County
Liaoning Province

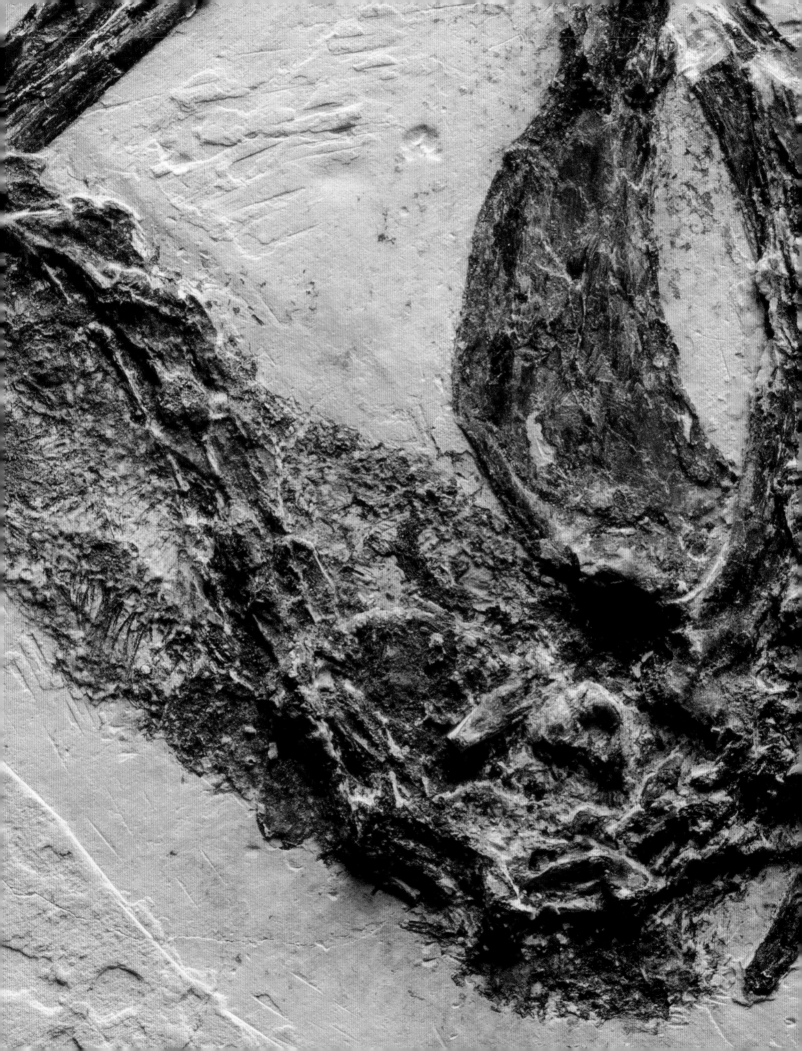

Birds have evolved an astonishing array of bill shapes and sizes, serving multiple purposes, from preening to building nests and from display to defense. However, the primary function of the bill is gathering food, and as such, the design of the jaws varies greatly depending on what the bird eats. Birds that probe the ground with their bills usually have long, slender jaws well adapted for sensing and grabbing the many invertebrates that live beneath the surface. The Jehol Biota includes birds—both enantiornithines and ornithuromorphs—which have beaks most likely specialized for probing the substrate along the muddy coast and in shallow waters of lakes. As in modern probers, the bills of these birds vary in length, suggesting that they foraged at varying depths, eating different foods while still living together.

Small enantiornithines such as *Rapaxavis pani* (*overleaf*) are unlikely to have penetrated the substrate much beneath its surface; the larger ornithuromorph *Xinghaiornis lini* (*left*) would have been able to reach to depths of 3 centimeters (1.2 inches), roughly the same probing depth of living oystercatchers. Probing pits and other foraging marks from Early Cretaceous track sites in South Korea (*below*) show clear evidence of how ancient birds took advantage of food resources available along mudflats and in shallow waters. Among these rarely preserved behaviors are marks left by birds foraging with the side-to-side swinging motion characteristic of present-day spoonbills (*below*). Evidence from Jehol fossils as well as contemporaneous track sites thus offer proof of the ancient origin of the probing behavior we see today among many birds living alongside bodies of water.

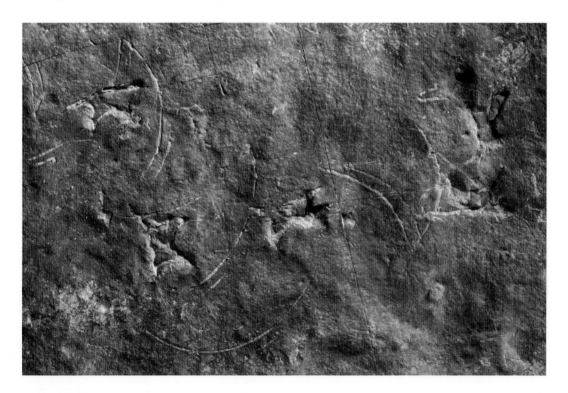

Xinghaiornis lini
XHPM-1121
Yixian Formation
Sihetun, Yixian County
Liaoning Province

Rapaxavis pani
DNHM-D2522
Jiufotang Formation
Lianhe, Chaoyang County
Liaoning Province

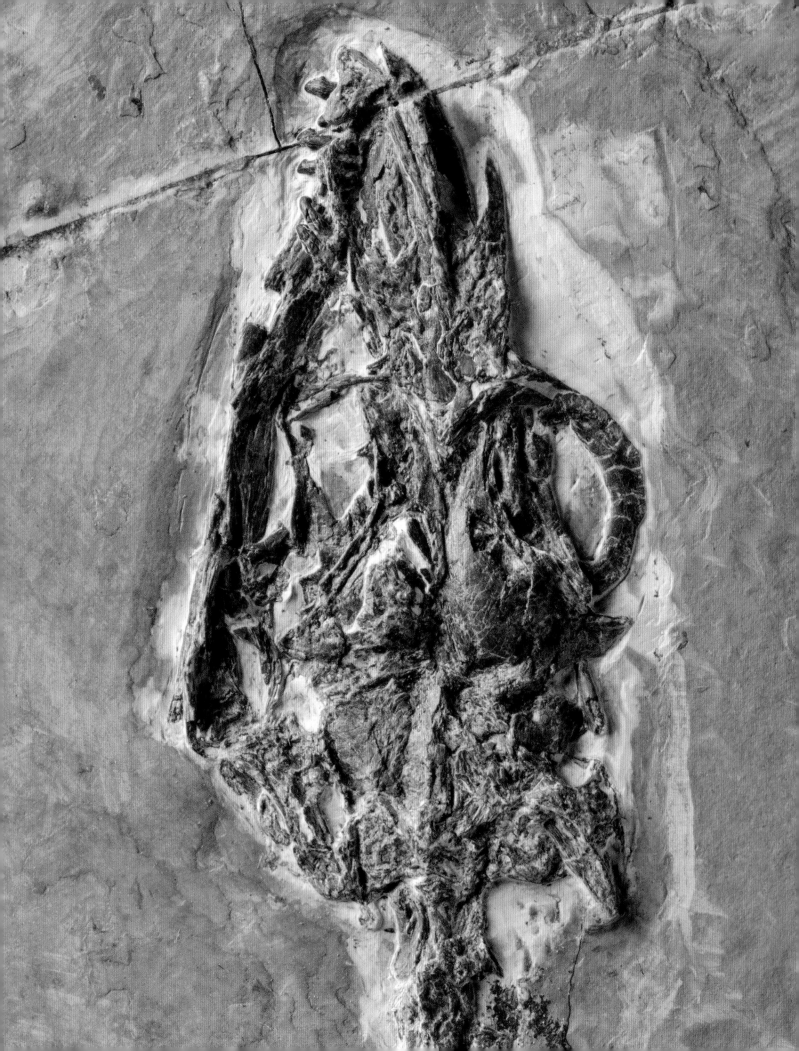

Most living birds are primarily active during the day, but some—including nightjars, whistling ducks, and others—are more active at night or near dusk. Whether early birds were more diurnal or nocturnal may be impossible to confirm, but studies looking at the eye orbits preserved in a number of Jehol fossils have shed some tentative light onto this question. The eye orbits of birds, either living or extinct, carry a ring formed by small bony plates (*below*). This scleral ring supports and maintains the integrity of the eyeball, and its size and relationship to the orbit gives us a notion of the animal's daily cycle. While more than one feature needs to be considered, for a given eye size, the internal opening of the scleral ring is typically smaller in diurnal birds than in their nocturnal counterparts. Thus, fossils preserving complete scleral rings together with well-preserved eye orbits allow comparisons between the diameters of these parts and those of living birds.

Unfortunately, not too many fossils possess well-preserved scleral rings in association with their eye sockets, but for some Jehol fossils, we have sufficient evidence to tell whether these birds preferred foraging under the sun or the stars. Such studies have shown that most Mesozoic birds, including *Confuciusornis sanctus*, *Sapeornis chaoyangensis* (*left*), and most enantiornithines (*overleaf*), were diurnal. These investigations, however, cannot speak to how well these birds saw or whether they had—like their living relatives—a broader visible spectrum than most mammals, including humans.

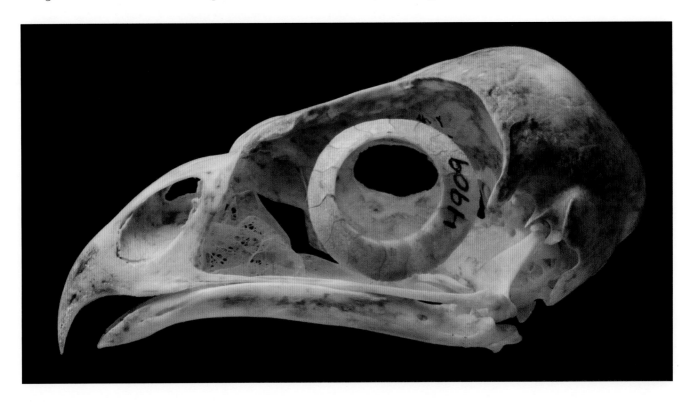

Sapeornis chaoyangensis
DNHM-D3078
Yixian Formation
Jianchang County
Liaoning Province

Longipteryx chaoyangensis
BMNHC-PH930B
Yixian or Jiufotang Formation
Lamadong, Jianchang County
Liaoning Province

Birds use their feet for different purposes. Herons and shorebirds have long legs and toes best suited for wading; ducks and gulls have webbed feet adapted for paddling; and birds of prey have powerful talons for seizing game. The variety of pedal adaptations of fossils helps us interpret what extinct birds did with their feet and the kinds of lifestyles they had. The grasping feet of the Jehol enantiornithines, with long and opposable hind toes ideally fit for clamping on branches and twigs (*left* and *below*), have consistently been used to identify these birds as denizens of the temperate forests that covered much of eastern Asia during the first half of the Cretaceous. The shape and overall proportions of the feet of these birds are very similar to those of modern perching birds, like most song-birds, in which the toes are designed to provide a powerful grip through a long and opposable hind toe, sharply curved claws on all toes, and elongate penultimate toe bones (*overleaf*). Fossils from the Jehol Biota also show that even the earliest enantiornithines that lived 131 million years ago—*Protopteryx fengningensis* and *Eopengornis martini*, among others—had already evolved these distinct adaptations for perching. These animals had relatively small sizes typical of living songbirds and relatively short, elliptical wings that allowed them to fly through even the thickest woods.

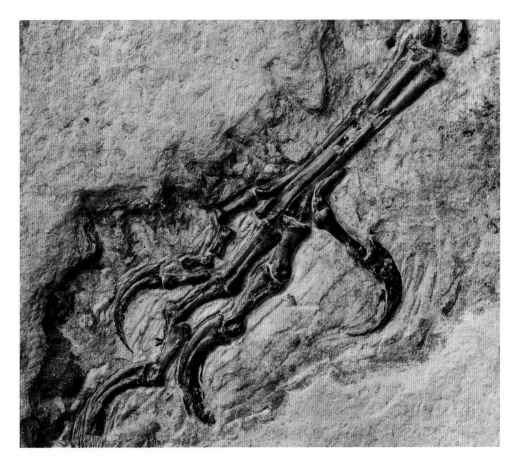

Enantiornithine indeterminate
BMNHC-PH1154A
Huajiying Formation
Sichakou, Fengning County
Hebei Province

Sulcavis geeorum
BMNHC-PH805
Yixian Formation
Lamadong, Jianchang County
Liaoning Province

Enantiornithine indeterminate
BMNHC-PH1068
Huajiying Formation
Sichakou, Fengning County
Hebei Province

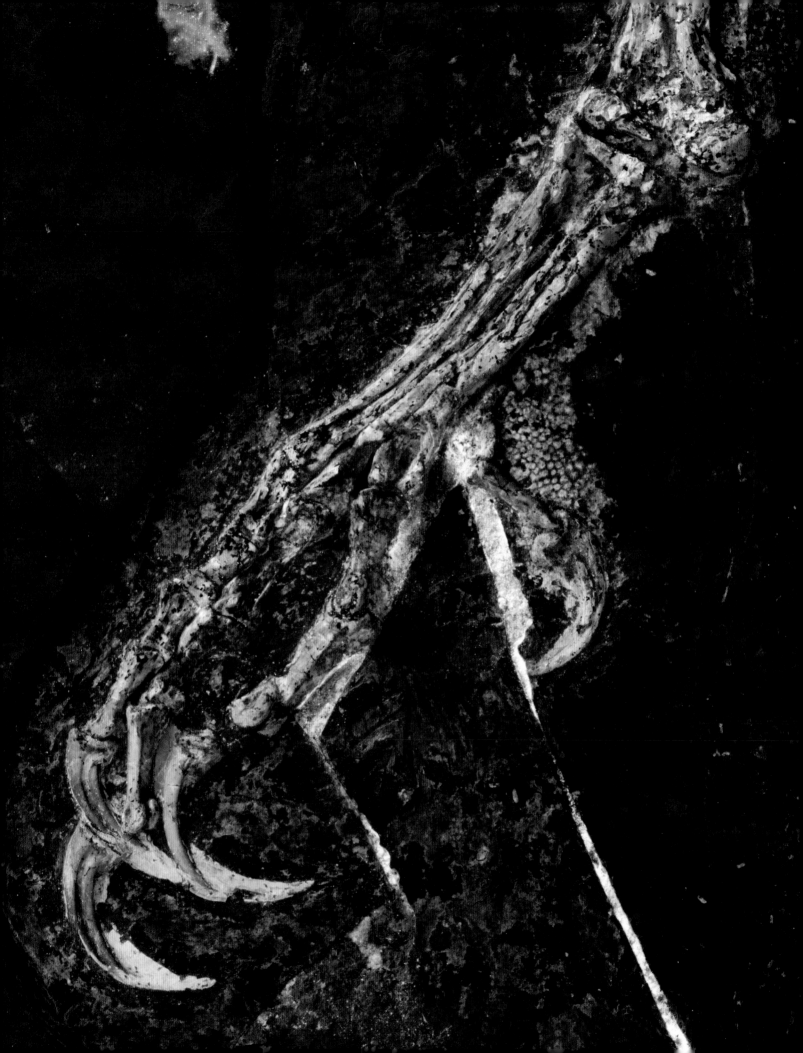

While it is easy to envision many enantiornithines dwelling in trees, flying and springing from one branch to another, other birds from the Jehol Biota are likely to have spent more time on the ground. The foot of the primitive short bony-tailed *Confuciusornis sanctus* (*left*) was unmistakably less adapted for perching than those of the Jehol enantiornithines. The opposable hind toe of *Confuciusornis* was much shorter in relation to other toes; the foot of this bird could not easily grip branches. Comparisons with living birds show that the proportions of the toes of *Confuciusornis* match those of pigeons and landfowl, birds that spend time both in trees and on the ground. Likewise, the feet of many primitive ornithuromorphs—*Archaeorhynchus spathula* (*below*), among others—also point to a more terrestrial existence. The feet of these birds are more strongly built and their toes are short and thick, ending in broad and less recurved claws. Moreover, the proportionally short hind toe of most early ornithuromorphs (fully reduced in *Archaeorhynchus*) was clearly ill suited for grasping. In general, the appearance of the feet of a number of these birds resembles those of landfowl, animals that spend a lot of their time walking and foraging on the ground. The perceived distinction in the lifestyles of many of these early ornithuromorphs with respect to their contemporaneous enantiornithines has lent support to the idea that these two kinds of primitive birds exploited the resources available in the Jehol environments very differently.

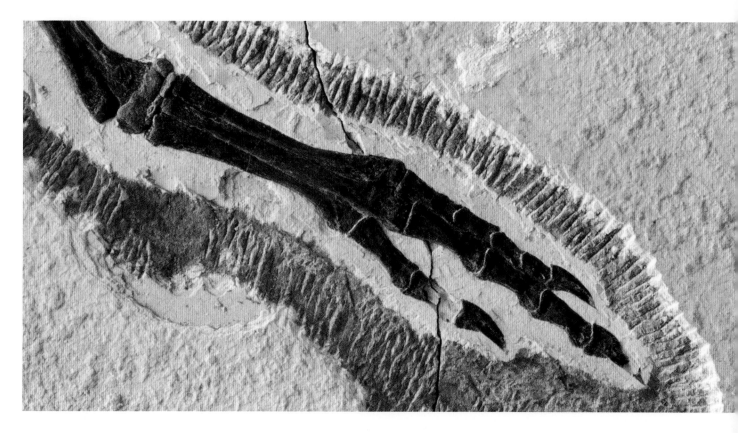

Confuciusornis sanctus
BMNHC-PH766
Yixian Formation
Sihetun, Yixian County
Liaoning Province

Archaeorhynchus spathula
IVPP-V17075
Jiufotang Formation
Jianchang County
Liaoning Province

The anatomy of the legs of hongshanornithids (*left*) and gansuids (*below*) tells us that a variety of primitive ornithuromorphs inhabited the coastal wetlands of the Jehol's ancient lakes. The elongate legs and delicate toes with small claws of *Hongshanornis longicresta* (*left*) are similar to those of shorebirds and other present-day birds that wade. Likewise, the similar configuration of the legs and feet of *Gansus zheni* (*below*) suggests that this larger bird was also a wader. The slender hip, short thighbone, and stacked foot bones of *Gansus*, features common to aquatic birds, indicate that it was also a skilled swimmer. The elevated hind toe of this bird would have barely touched the ground and the three main toes might have been joined by some degree of webbing, as is common of living birds with elevated hind toes. These adaptations speak of amphibious birds that waded and swam near the coastline.

Coeval footprint sites in South Korea provide direct evidence of such behavior. They show an abundance of web-footed tracks of different sizes, as well as a variety of feeding behaviors, indicating that many different kinds of birds waded in shallow water and took advantage of resources offered by wetlands that during the Early Cretaceous dotted what is now the Korean Peninsula and the neighboring northeastern corner of China. The feet of Jehol birds such as *Gansus* and *Hongshanornis*, together with the evidence available from contemporaneous track sites, tell us that wading and foraging on these environments were common among some of the earliest relatives of today's birds.

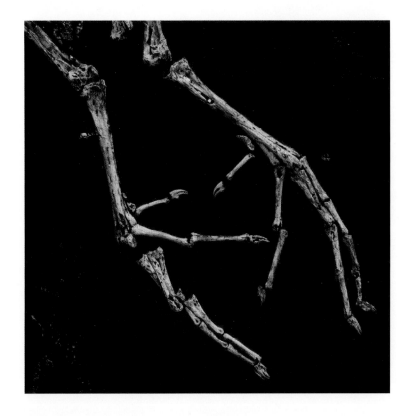

Hongshanornis longicresta	*Gansus zheni*	*Gansus zheni*
DNHM-D2945	BMNHC-PH1342	BMNHC-PH1392
Yixian Formation	Jiufotang Formation	Jiufotang Formation
Lingyuan County	Sihedang, Lingyuan County	Sihedang, Lingyuan County
Liaoning Province	Liaoning Province	Liaoning Province

BMNHC-Ph001392.

The Jehol Biota's outstanding fossil collection has given us the most significant evidence for documenting the key transformations that took place during one of the most captivating evolutionary transitions in the history of life: from the ground-dwelling dinosaurs that characterize the Mesozoic Era to the airborne birds familiar to all present-day environments. This remarkable paleontological bonanza has enabled us to create a vivid portrait of the lives of many birds that thrived in the bygone world of the Mesozoic dinosaurs. The exquisite preservation of these fossils (*right*) has allowed us to reconstruct even the finest anatomical features of some of the oldest and most distant relatives of living birds, and a wealth of details pertaining to their ecology, development, and flight performance.

These fossils show that birds diversified into a variety of ecological niches—arboreal, terrestrial, and aquatic—very early in their history; that the elaborate anatomical, physiological, and behavioral properties of living birds—adaptations linked to their reproduction, digestion, growth, sexual displays, and locomotion—appeared first among different primitive groups (and, in many instances, surprisingly early); and that the stunning aerial competence of today's birds developed as a result of a complicated succession of fundamental adaptations that evolved over many millions of years. Since the time of Charles Darwin, scientific studies of birds have provided paramount information for understanding a wealth of questions related to the evolution of vertebrate animals. Studies of the exceptionally preserved, abundant, and highly diverse Jehol avifauna has enabled us to extend this century-old legacy into the paleontological realm, enlightening a suite of evolutionary phenomena related to animals that lived long ago. Like no other fossils, the spectacular avifauna from the Jehol Biota has brightened our understanding of the lives of a thriving diversity of ancient birds, which study has transformed our knowledge about some of the earliest relatives of present-day birds and has greatly clarified key aspects of the evolution of these remarkable animals.

Confuciusornis sanctus
HGM-41HIII0400
Yixian Formation
Liaoning Province

THE EARLY EVOLUTION *of* BIRDS

The stunning diversity of birds from the Jehol Biota developed during the second half of the Mesozoic Era, a time interval often referred to as the Age of Dinosaurs. Spanning approximately 252 to 66 million years ago, the Mesozoic is subdivided into the Triassic, Jurassic, and Cretaceous Periods; the boundaries between them are defined by significant turnovers in the composition of the fauna and flora that characterized each of these geological periods. The Mesozoic Era was predated by the Paleozoic Era (roughly 541 to 252 million years ago)—a time that witnessed the dawn of most major groups of organisms, from mollusks and vertebrates to ferns and conifers—and was followed by the Cenozoic Era (approximately the past 66 million years of Earth's history), in which the species that live today had their origins. The first birds originated in the Mesozoic—the earliest record being that of the renowned *Archaeopteryx lithographica*, some 150 million years ago—and the Jehol Biota of northeastern China flourished during the first half of the Cretaceous, from about 131 to 120 million years ago.

The Mesozoic world was a very different place from the one we inhabit today. Average temperatures were much warmer, their latitudinal gradients were significantly weaker, and the polar regions were rarely covered by permanent icecaps. The configuration of the continents and their relative positions were also considerably different. In the Triassic (252 to 201 million years ago), all landmasses were clustered into one enormous supercontinent called Pangaea. This configuration greatly influenced the climate and many environmental features of this colossal landmass. Rocks from around the world indicate that the Triassic environments were predominantly hot and dry. In what was perhaps the hottest period in the history of Earth, deserts spread out over much of Pangaea's interior regions, particularly in the mid- and low latitudinal regions. In this enormous continent, the latitudinal differences in temperature and rainfall were not as noticeable as those of today, although the seasons were strongly marked by scorching summers and cold winters. Toward the second half of the Triassic, the very same tectonic forces (the geologic processes that have shaped the Earth's crust for eons) that had previously consolidated all landmasses into one supercontinent began to break Pangaea apart. Such gradual fragmentation was followed by changes in the global climate, which became gradually more humid, even though the rocks of that time still document the presence of large deserts north and south of the equator, and rapidly fluctuating conditions at low latitudes.

As the Triassic gave way to the Jurassic (201 to 145 million years ago), persistent tectonic forces kept separating Pangaea in half. Such changes in the configuration of the landmasses, and the consistent trend toward global humidification, led to major monsoons that swept across extensive regions. The sea levels also rose significantly, flooding the margins and, in some instances, even the interior of the landmasses. As the climate became increasingly moist, the deserts that covered much of Pangaea during the Triassic began to give way to tropical and temperate forests. By the middle of the Jurassic, Pangaea had fragmented into two large continents, Laurasia to the north and Gondwana to the south, separated by an equatorial sea. The fragmentation of Pangaea continued as the Jurassic transitioned into the Cretaceous (approximately 146 to 66 million years ago). The climate around this transition, however, became cooler, an anomaly compared to the elevated temperatures typical of the "greenhouse" world of the Mesozoic. There is evidence that around this time the average temperatures of temperate regions, such as those occupied by the Jehol Biota of northeastern China, might have been similar to those of today and that glaciation events might have blan-

Late Cretaceous   90 Million Years Ago

Early Cretaceous   120 Million Years Ago

Jehol Biota

Middle Jurassic   170 Million Years Ago

Early Jurassic   200 Million Years Ago

Early to Middle Triassic   240 Million Years Ago

The geographical evolution of landmasses during much of the Mesozoic brought about a dramatic change in the configuration of the continents, from a single Triassic landmass—Pangaea—to a continental distribution resembling that of today. The red dot indicates the approximate position of the Jehol Biota as it existed in the Early Cretaceous, between 131 and 120 million years ago.

keted the poles with ice. The cool weather that prevailed during the first portion of the Cretaceous warmed with the increased tectonic activity that led to the early fragmentation of Laurasia and Gondwana, and the opening of the Atlantic and Indian Oceans, respectively.

The extensive fragmentation of these two continental landmasses during the Jurassic–Cretaceous transition was associated with widespread volcanic eruptions that led to a rise in carbon dioxide levels. This extensive period of volcanic activity would play a key role in the spectacular preservation of the Jehol Biota. By the middle of the Cretaceous, some 100 million years ago, the rising levels of carbon dioxide brought global temperatures back to the high levels of previous Mesozoic times. As Laurasia and Gondwana became more fragmented, warm seas flooded large portions of land, covering one-third of the present-day continents by the Middle Cretaceous. As the emerging lands spread over a much wider latitudinal range, climatic zonation increased and the polar regions experienced episodic glaciations. Yet the climate continued to be characterized by less latitudinal gradients than those of today, which allowed the development of polar forests that hosted a wide diversity of organisms. By the end of the Cretaceous, the continental configuration was approaching the general appearance of many of today's landmasses. Vast portions of the continents were still under shallow seas, India stood as a large island at the center of the Indian Ocean, and Australia and South America remained connected to Antarctica. This was the appearance of the world when a 10-kilometer (a little more than 6 miles) asteroid plunged into the ancient Caribbean Sea, causing global havoc and bringing the Mesozoic Era to a close.

MESOZOIC LIFE The climate and the continental configuration were not the only ways in which the world of the Mesozoic was different from that of today. The plants and animals that lived during this time were also quite unlike today's, and even if many modern groups of organisms originated during the Mesozoic, their relative abundance was different as well. The Triassic followed the largest mass extinction our planet has ever endured, occurring at the end of the Paleozoic Era (some 252 million years ago), which scientists estimate wiped out more than 90% of all that was alive at the time. This period of intense biological turnover witnessed the rise to dominance of many groups of plants and animals. Seed-bearing plants, such as extinct types of seed-ferns, cycads, ginkgoes and their relatives, and a variety of conifers, came to dominate the various landscapes. Giant, distant cousins of modern amphibians, which superficially resembled supersized versions of Chinese giant salamanders, thrived in the streams and shallow lakes. A variety of reptiles known as synapsids, which include the ancestral stock of mammals, colonized many of the arid regions of Pangaea, and another group of reptiles, the archosaurs, ascended to dominance in numerous terrestrial environments. Many different groups of archosaurs, including various types of crocodile-like animals, winged pterosaurs, and dinosaurs, evolved during the Triassic.

The earliest dinosaurs appear in the fossil record about 230 million years ago. These animals were small, bipedal predators and omnivores that shared a variety of environments with other Triassic reptiles. Dinosaurs steadily diversified during the Triassic, separating into several main groups: the omnivorous and/or herbivorous ornithischians and sauropods, and the carnivorous theropods. The latter would in time give rise to the earliest birds, which would extend the pedigree of these fearsome animals into the present.

The precise reasons for the diversification that dinosaurs experienced toward the end of the Triassic, particularly in mid-latitudes, are not entirely understood. A variety of factors might have played a role in the extinction of many groups of synapsid and archosaur reptiles that shared the land with the earliest dinosaurs during this period. The fossil record, however, clearly shows that the last 30 million years of the Triassic saw the emergence of dinosaurs, which progressively came to dominate the scene in all terrestrial ecosystems.

The continental fragmentation that characterized the Jurassic intensified the already ongoing evolutionary diversification of the dinosaurs. During this period, stegosaurs, ankylosaurs, brachiosaurs, allosaurs, coelurosaurs, and many other familiar types of dinosaurs made their debut. Birds also originated during the Jurassic from within a group of small theropod dinosaurs called maniraptorans. The evolutionary transition from these early maniraptorans to birds—first represented by the famous *Archaeopteryx* of the latest Jurassic—has been beautifully documented by discoveries of the past 25 years, many of which occurred in China. The wetter climate of the Jurassic stimulated the development of extensive mixed forests formed by conifers (yews, monkey puzzle trees, and other cone-bearing trees), ginkgoes, cycads, and ferns. This abundant vegetation sustained a growing diversity of herbivorous dinosaurs, which evolved into colossal sizes, particularly among the long-necked sauropods. Sauropods shared these food resources with various other herbivorous dinosaurs such as the plated stegosaurs and a number of smaller and lighter forms (heterodontosaurs, dryosaurs, and iguanodonts), which were equally preyed on by different types of carnivores.

By the Late Jurassic, pterosaurs—the flying reptiles that had originated in the Triassic—had evolved into numerous new types occupying a variety of aerial niches. Throughout the second half of the Mesozoic, these majestic animals shared the sky with the earliest birds and a diversity of other bird-like dinosaurs (*Anchiornis huxleyi* and *Pedopenna daohugouensis*, among others), the anatomy and feathering of which indicate that they were capable of flight. Having originated in the Triassic, mammals also evolved different lifestyles in the Jurassic, although these animals remained small throughout the Mesozoic Era. As had become the case in the Triassic, the warm waters of the Jurassic oceans were home to a wide range of large marine reptiles and a wealth of other organisms. The dolphin-like ichthyosaurs that had originated in the Triassic became highly diverse, and the terrifying plesiosaurs made their first appearance. Many of these sea reptiles lived in tropical reefs (frequently formed by mollusks called rudists), which supported an abundance of diverse fish and invertebrates.

Fossils from the Cretaceous show a significant step toward the modernization of the fauna and flora. Many major groups of modern animals, including different types of insects, crocodiles, and mammals, first appeared during the Cretaceous. Perhaps the most significant event of this period is the origin and diversification of flowering plants, which brought the Mesozoic ecosystems closer to those of today. The first of these plants appeared early in the Cretaceous and rapidly underwent great diversification; some of their earliest representatives are known from the Jehol Biota. Not surprisingly, the fossil record shows how the dawn of flowering plants coincided with the first appearance of many modern groups of insects, especially those that include pollinating species. Fossils of these early pollinators are still rare, but recent genetic studies have estimated the timing of the spectacular diversification of the major groups of pollinating insects—bees, wasps, flies, and butterflies—to the Early Cretaceous, during which the

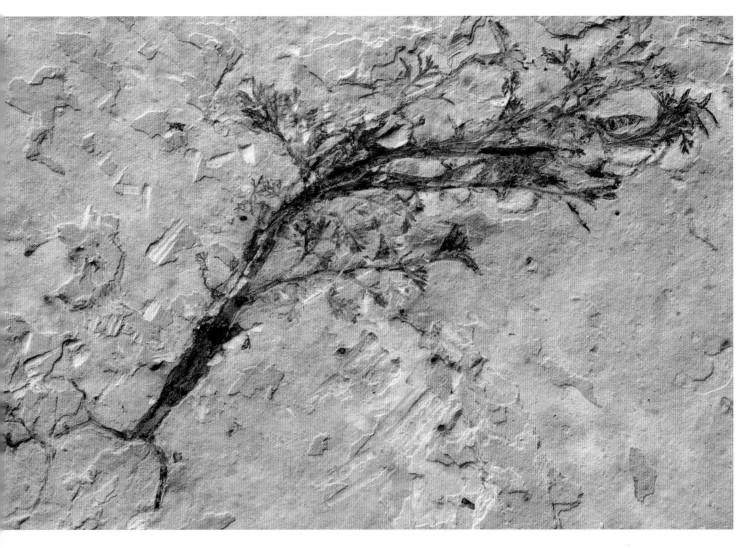

diversification of the earliest flowering plants also occurred. Later in the Cretaceous, the herbaceous flowering plants that first appeared at the beginning of this period would evolve into the array of thriving trees and other plants that dominate our planet today, and the interdependent nature of flowers and insects would be fully established.

Many groups of dinosaurs also flourished during the Cretaceous. In addition to iconic groups such as the fearsome tyrannosaurs, the horned dinosaurs, the armored ankylosaurs, and the duck-billed dinosaurs, the Cretaceous witnessed the diversification of numerous types of small, bird-like dinosaurs. Birds also underwent a tremendous evolutionary radiation and so did the pterosaurs, which evolved into the largest animals to ever take flight into Earth's skies. The extensive Cretaceous seas also teemed with life. Ichthyosaurs became extinct early in this period, but enormous plesiosaurs continued to lurk in the waters that also became home to the mosasaurs, another group of gigantic predatory reptiles highly specialized for life in the ocean. Fish of all sizes and a myriad of invertebrate animals—the shelled mollusks called ammonites, among the most notable—roamed the warm Cretaceous seas, which were also inhabited by the 1.2-meter-long (4 feet) *Hesperornis regalis* and its kin of foot-propelled diving birds.

The abundant rocks that formed during the Mesozoic Era have left us with a vivid picture of the ancient world that was home to many dinosaurs and countless other

The delicate fossils of the aquatic *Archaeophructus sinensis* represent some of the earliest known examples of flowering plants. Fields of this small plant might have lined the shores of the ancient Jehol lakes. The origin and rise to dominance of flowering plants was one of the most significant biological events of the Cretaceous.

organisms that lived alongside them. It was a time in which the shape of the continents, the distribution of the oceans, the characteristics of the climate, and the variety of the animals and plants that inhabited the marine and terrestrial habitats were very different from what they are today. Among the most notable events that took place in this utterly different world was the evolutionary transition of a group of fearsome carnivorous dinosaurs into the feathered and acrobatic animals that we call birds.

THE RISE OF BIRDS All living birds are the descendants of ancestors that first evolved more than 150 million years ago; these familiar animals are the product of millions of years of stepwise evolution. For many decades, the search for their origin and their closest relatives was at the center of an intense controversy, but the discovery of numerous recent fossils, particularly from the Jehol Biota, has largely settled the issue. Already in the nineteenth century, scientists recognized the similarity in the shapes and features of numerous bones between the handful of carnivorous theropod dinosaurs known at that time, the Late Jurassic *Archaeopteryx*—whose earliest fossil discoveries were studied in the 1860s—and modern birds. Thomas Henry Huxley, the famous British anatomist and defender of Charles Darwin's views of evolution, highlighted 35 features in common between the large theropod *Megalosaurus bucklandii* and birds. Yet, the idea of an evolutionary connection between birds and certain carnivorous dinosaurs tumbled around the turn of the twentieth century, and during much of that century, theropod dinosaurs were often regarded as too specialized to be the ancestral stock of birds.

This view, however, started to change in the 1960s and 1970s, when new discoveries of the bird-like dromaeosaurid theropod *Deinonychus antirrhopus* and more detailed studies of *Archaeopteryx* highlighted once again the resemblance between the skeletons of these animals, on the one hand, and those of living birds, on the other. Such studies, primarily those by the Yale University paleontologist John Ostrom, brought the idea that birds originated from an ancestral theropod dinosaur back into the limelight. Since then, a great deal of new evidence in support of such a notion has been amassed. Today, most scientists agree that birds are evolutionarily nested within a particular group of theropod dinosaurs—the maniraptorans—which includes the large-headed, sickle-clawed dromaeosaurids (*Deinonychus* and its kin), the small-toothed, lightly built troodontids (*Troodon formosus* and *Mei long*, among others), and the parrot-headed, short-tailed oviraptorids (*Oviraptor philoceratops* and relatives), among several other groups of relatively bird-like dinosaurs. Recent discoveries, and new analyses, have revealed that these maniraptoran theropods share many anatomical, physiological, and behavioral features with birds and that numerous characteristics previously thought to be exclusively avian first appeared among these dinosaurs.

Studies from the past few decades have also shown that the skeletons of birds and other groups of maniraptoran dinosaurs share a number of features that had already evolved in the earliest dinosaurs and their closest relatives, animals that lived in the Triassic, more than 230 million years ago. Many of these anatomical trademarks are related to how small, early predatory dinosaurs stood upright and moved. These discoveries indicate that even the earliest dinosaurs held their hindlimbs in a vertical position directly beneath their bodies and walked on two legs, the same bipedal type of locomotion that we see today among birds; even details of the leg musculature of these early dinosaurs have been shown to be similar to those of living birds. There is little doubt that the transformation of the hip bones and the rearrangement of the hindlimbs that allowed the closest relatives of dinosaurs to adopt an upright posture gave the earliest dinosaurs an edge over other Triassic reptiles, and that it paved the road for dinosaur dominance during the rest of the Mesozoic. Bipedalism was also critical for the evolution of birds and the development of flight because by evolving such a novel type of locomotion, the forelimbs became free to grow into the wings that propelled the first of these animals into the air.

The fossil record also tells us that birds not only inherited the bipedalism of the earliest dinosaurs but also their characteristic air-filled bones, wishbones, three forward-facing toes, and S-shaped necks, all of which evolved among some of the earliest theropods. Fossil discoveries also show that many more skeletal features of birds are found among dromaeosaurids, troodontids, and other groups of maniraptoran theropod dinosaurs. Salient among these features are the long forelimbs that support the feathered wings, large breastbones that today anchor the muscles that power the flight stroke, and semicircular wrist bones that allow the tips of the wings to swivel during flight and the entire wings to fold against the sides of the body. The anatomy of these early maniraptoran relatives of birds also hints at the presence of vital organs that are unique to birds among living animals, including their flow-through lung system with associated air sacs and enlarged brains with structures specialized for visual and sensory integration. Paleontological breakthroughs of the past few decades have shown that the characteristics that are shared between birds and the different types of early maniraptoran dinosaurs are distributed throughout the skeleton, and that many features thought to be restricted to the bird skeleton have now been found in the fossils of these and other theropod dinosaurs.

Other discoveries reveal that the nesting behavior of certain groups of early maniraptoran dinosaurs was like that of living birds. Fossils of oviraptorids and troodontids, unearthed on top of their egg clutches, reveal that these animals adopted postures comparable to those of brooding birds. These fossils have documented that the 2.5-meter-long (a little more than 8 feet) *Troodon formosus* and its lightly built relatives sat directly on top of vertically oriented and partially buried eggs with their legs tucked beneath their bodies. While nesting, the bodies of these animals would have contacted the exposed portion of the eggs, possibly incubating them at temperatures higher than those of the environment. In the heavier oviraptorids, the available fossils show that the

The 230-million-year-old *Herrerasaurus* is one of the most primitive theropod dinosaurs. The skeleton of this 3-meter-long (almost 10 feet) animal illustrates the upright posture and bipedal gait that we still see among its living avian descendants.

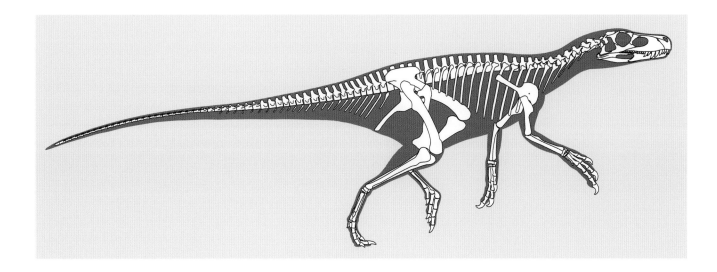

adult tucked its legs within a space at the center of the egg clutch and hugged the clutch with its long arms. These discoveries uphold the notion that, like modern-day birds, early maniraptoran dinosaurs protected their eggs for a prolonged period of time.

A wealth of further evidence indicates additional resemblances to living birds in the characteristics of the eggshell and the shape of the eggs. Like all other dinosaurs, as well as living crocodiles and birds, early maniraptorans laid hard-shelled eggs in which the shell was made of the carbon calcium mineral calcite. In crocodiles and most Mesozoic dinosaurs, the eggshell is formed by a single layer of calcite crystals, but in birds and maniraptoran dinosaurs such as dromaeosaurids, troodontids, and oviraptorids, the eggshell exhibits more layers; the calcite crystals are typically arranged in two to three distinct layers stacked on top of each other. In birds and these maniraptoran dinosaurs, the eggshell is also pierced by a fewer number of air holes (which allow the developing embryos inside the eggs to breathe) than those in the eggs of crocodiles and other groups of Mesozoic dinosaurs. Likewise, while the eggs of many Mesozoic dinosaurs are nearly symmetrical—they are either round or their two poles are very similar in shape—those of early maniraptoran exhibit the lopsided shape characteristic of avian eggs. The eggs of early maniraptorans also have volumes that are bigger, with respect to the adult body size, than those of other Mesozoic dinosaurs. In fact, the relative size of the eggs of early maniraptoran dinosaurs approaches the high egg-volume to body-size ratio typical of living birds.

Further paleontological evidence documents physiological attributes of the reproduction of early maniraptoran dinosaurs that are uniquely shared with birds. The discovery of a gravid oviraptorid female, fossilized with a pair of shelled eggs inside her pelvic canal, provides direct confirmation that these maniraptoran dinosaurs laid two eggs at a time and at discrete time intervals. Every other kind of reptile, including other dinosaurs, lays all their eggs at once. Birds, however, lay their eggs at discrete intervals of time, often one a day, and it takes them days to amass an entire clutch. Modern birds have a single functional oviduct—typically the left one—and hence, they can only ovulate one egg at a time. Nevertheless, their more primitive maniraptoran relatives, like the exceptional oviraptorid female fossilized with two eggs about to be laid, had two functional oviducts and the ability to lay two eggs at the same time. Fossil evidence suggests that it also took a prolonged period of time—perhaps up to several days—for these dinosaurs to lay their entire egg clutches. Such an interpretation is based on the discovery of dozens of egg clutches of different types of Mesozoic theropod dinosaurs that show a distinct pattern of egg-pairing, which in turn suggests that they were laid a pair at a time. All this evidence, amassed from around the world and including dinosaurs that lived in different geologic periods, strongly indicates that key components of the reproductive system of living birds had already evolved among their maniraptoran predecessors.

More clues about the ancestry of birds have come from the characteristics of the fossilized bone tissues preserved in a variety of Mesozoic dinosaurs. The study of these tissues, which often maintain their cellular microstructure in exquisite detail, allows us to determine the growth rates of these dinosaurs and other extinct animals. These investigations have documented that while once believed to be slow-growing behemoths, the extinct dinosaurs from the Mesozoic Era generally grew much faster than typical living reptiles (crocodiles, turtles, and lizards). Growth patterns among these dinosaurs have also been shown to be directly proportional to their size—the larger the animal, the faster it grew. While smaller dinosaurs grew at rates slower than those of living birds (but still faster than those of living crocodiles, turtles, and lizards), recent studies of the bone tissue of small

A clutch of 12-centimeter-long (nearly 5 inches) eggs from a Late Cretaceous oviraptorid dinosaur shows the typical paired arrangement of the eggs of many early maniraptorans. Such spatial organization suggests that these dinosaurs laid sets of two eggs over the span of hours or even days.

maniraptorans, like the 70-centimeter-long (28 inches) dromaeosaurid *Mahakala omnogovae* have shown that these dinosaurs grew at rates comparable to those of *Archaeopteryx* and other early birds. The correspondence between the bone tissues of *Mahakala* and *Archaeopteryx* indicates that the most primitive birds and their maniraptoran forerunners were physiologically similar when it came to growth strategies.

Furthermore, special types of bone tissues, such as the medullary bone characteristic of ovulating birds, have been documented in the females of certain extinct theropods, such as *Tyrannosaurus rex*. Medullary bone is a highly porous, well-vascularized, and calcium-rich type of tissue that functions as a calcium reservoir inside the long bones of female birds during their breeding season. This tissue is only present during the egg-laying season—it begins to form before the bird starts laying eggs, and it is finally resorbed days after the egg-laying season has ended. The discovery of medullary bone inside the long bones of extinct theropod dinosaurs underscores the extent to which the reproductive strategies of living birds have been inherited from their dinosaurian predecessors.

Fossil discoveries have also shown that the gargantuan sizes stereotypically associated with the dinosaurs of the Mesozoic Era were not common to all these extinct animals. In fact, a growing number of fossils indicate that many theropod dinosaurs were small when full-grown and that the gigantism characteristic of many groups of Mesozoic dinosaurs evolved independently from animals of much more modest dimensions. Recent studies have provided evidence that a long evolutionary trend of miniaturization—toward body sizes comparable to those of many living birds—predated the origin of birds by many millions of years. Today we know of different groups of theropod dinosaurs—from the Triassic to the Cretaceous and across the entire family tree of these animals—which included representatives whose sizes barely exceed those of many present-day birds. This evidence shows once again that, like numerous other characteristics of modern birds, the small sizes of these animals were inherited from their dinosaurian predecessors.

Not surprisingly, the many common attributes shared by living birds and the different types of extinct theropod dinosaurs can be extended to the genetic toolkit that underpins the appearance, behavior, and physiology of all organisms. Modern techniques have figured out how to extract portions of DNA, as well as entire genomes (the complete set of genetic material of an organism), from the bones and other preserved tissues of numerous ancient animals. Using these developments, scientists have been able to study the genetic material of mammoths, dire wolves, and even our Neanderthal cousins; yet no one has ever been able to recover unquestionable DNA—let along a genome—from the fossils of dinosaurs that lived during the Mesozoic Era. However, the well-known relation between the size of the cells of an organism and the size of its genome has allowed us to peek into the characteristics of the genomes of these extinct dinosaurs; applying this relation to the bone cells and genomes of modern animals, one study was able to estimate the genome sizes of Mesozoic dinosaurs.

Simply put, this study measured the spaces that used to house bone cells, visible in paper-thin sections of dinosaur bones, and applied the cell-to-genome size ratio of living animals to calculate the sizes of the genomes of duck-billed, long-necked, and other extinct dinosaurs, including dromaeosaurid maniraptorans. Interestingly, the majority of the theropod bones sampled by this study pointed at genomes significantly smaller than those estimated for other Mesozoic dinosaurs and comparable to those of living birds, which have genomes that are usually half the size of those of mammals and reptiles such as crocodiles. Recent molecular studies have identified a suite of genetic processes, from the disappearance of repetitive portions of DNA to the loss of entire genes, as responsible for the small genomes of present-day birds. The realization that the mighty *Tyrannosaurus rex* and the formidable dromaeosaurids had genome sizes comparable to those of living birds tells us that these genetic processes were at play millions of years ago among the theropod forerunners of birds. Once again, this evidence underscores the diversity of attributes that were formerly assumed to be characteristically avian but have now been determined for the dinosaurs that most scientists identify as being closer to the origin of birds.

Despite the numerous and diverse lines of evidence that today support the notion that birds are evolutionarily nested within theropod dinosaurs, nothing has consolidated this idea more than the discovery of feathers—the quintessential bird feature—in a variety of early maniraptoran dinosaurs. Most of these exceptional fossils are from the Jehol Biota, but evidence of the plumage that covered the bodies of different kinds of theropods has also been unearthed in North America, Europe, and other places in Asia. The feathery coverings of these animals varied greatly, but the fact that many kinds of extinct, feathered dinosaurs have been discovered indicates that these structures were widespread among the theropod predecessors of birds. Additionally, tantalizing skin structures found in association with both pterosaurs and some ornithischian dinosaurs have prompted the claim that feathers might have originated prior to the rise of theropod dinosaurs and in animals far removed from the ancestry of birds. The interpretation of these skin structures is controversial, but their existence suggests that the evolutionary history of feathers is conceivably more complex than previously thought. The fossil record provides undeniable evidence that the closest maniraptoran relatives of birds had feathers of modern appearance.

The wealth of fossil evidence unearthed in the past few decades has revealed how many of the features previously thought to be exclusively avian—from feathers to wishbones and from downsized bodies to small genomes—are now known to have evolved, over many millions of years, among the theropod predecessors of birds. The degree of birdness illustrated by the different types of early maniraptoran dinosaurs discovered during these decades has blurred the line between birds and their immediate predecessors. Undoubtedly, these fossil discoveries have taught us a lot, but the picture is still very complex.

Our growing understanding of the evolution from early maniraptorans to birds has also revealed how many of the features involved in this multimillion-year-old transition have a more convoluted history than what was previously anticipated. Many features and structures (in the brain, the beak, the feathers, and elsewhere) that were thought to have developed gradually and to have become modernized continually are now known to have experienced a significant degree of evolutionary convergence, having evolved independently in a number of lineages along the long journey from these theropod antecessors to birds. All the different lines of evidence available today indicate that modern birds are just a branch of an extensive family tree that contains not only the many kinds of birds that lived during the Mesozoic Era but all the dinosaurs that

203

Numerous studies of the evolutionary relationships of dinosaurs support the inclusion of birds within a group of maniraptoran theropod dinosaurs called paravians. These also include the familiar *Velociraptor* and many other feathered dinosaurs that are considered to be the closest relatives of birds.

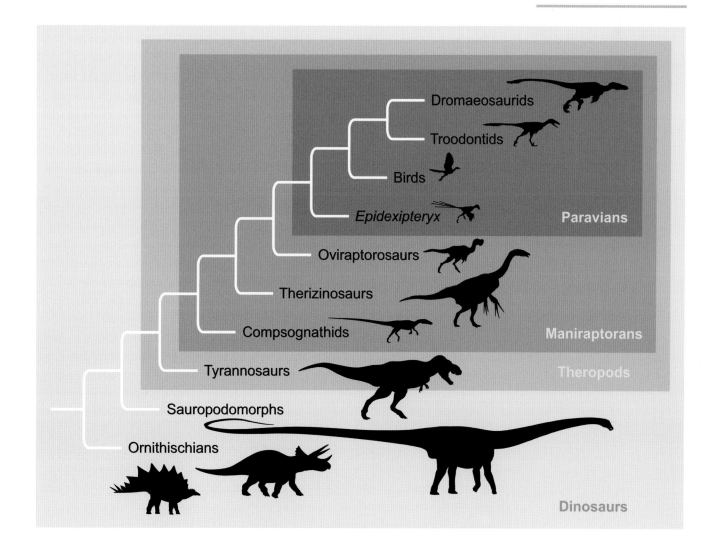

reigned on Earth for many millions of years. To emphasize birds' membership in the dinosaur clan, scientists refer to them as avian dinosaurs and to the many groups of their extinct Mesozoic predecessors as non-avian dinosaurs.

One hundred and fifty years after the earliest arguments in support of an evolutionary connection between birds and non-avian dinosaurs were proposed, there is a universal understanding that the amazing animals that are icons of the Mesozoic Era are not entirely extinct. Certainly, the famous *Triceratops horridus* and *Tyrannosaurus rex* vanished when the Age of Dinosaurs came to an end, some 66 million years ago, but the early ancestors of some lineages of living birds survived the same mass extinction that drove these celebrated dinosaurs to their demise. As heirs of the dinosaurs that had given them origin millions of years earlier, these early forerunners of modern birds diversified, evolving into the colorful feathered cohort that we see today.

FEATHERED DINOSAURS Feathers are an essential and universal feature of birds. It is thus not surprising that the discovery of feathers in fossils that would otherwise be unquestionably classified as theropods has greatly strengthened the idea that birds evolved from these dinosaurs. The first indication that feathers covered the bodies of some non-avian dinosaurs came in 1996, when a complete, 60-centimeters-long (24 inches) skeleton of the theropod *Sinosauropteryx prima* was exhumed from 125-million-year-old rocks near the rural village of Sihetun, in the western portion of Liaoning Province (northeastern China). The skeleton of this short-armed and long-tailed dinosaur was lined by a blackish halo of filament-like structures, preserved on the slab containing the fossil, which gave the animal a fuzzy appearance. The spectacular preservation of this fossil showed that the length of these soft filaments varied substantially across the body, as was also the case for the space between the base of these filaments and the bones they covered. Around the skull and the hip, the bases of the filaments were next to the bones, but around the back of the neck and tail these structures were widely separated from the skeleton. Such a pattern clearly reflected the amount of flesh—muscles, fat, and other soft tissues—separating the skin (and the bases of the fuzzy structures) from the bones in different parts of the body. This observation and the overall appearance of these filaments indicated that these structures projected outward from the skin surface and covered the entire body of this ancient dinosaur. Additionally, microscopic observations suggested that the bases of the main filaments were hollow and that thinner filaments branched off from these bases, thus overall resembling modern downy feathers.

Since the discovery of *Sinosauropteryx*, hundreds of fossils representing more than 20 species of theropod dinosaurs with feathery coverings have been found in the Jehol Biota of Liaoning Province and other Mesozoic sites in northeastern China. Yet these downy dinosaurs were living around the globe. Evidence indicating the presence of plumage in non-avian theropods has also been recovered from fossils found in Europe, North America, and other parts of Asia. These species are spread over a large portion of the family tree of theropods, and the appearance of their plumage varies greatly, from simple or loosely branched filaments often referred to as protofeathers, to highly organized structures with a shaft, vanes, and all the attributes of a modern feather. The fact that these dinosaurs represent different branching events in the long evolutionary saga of these animals suggests that the extraordinary discoveries of the past two decades are just scratching the surface of a much greater and still undiscovered diversity of feathered dinosaurs.

New feathered dinosaurs are being added to the staggering roster of Jehol fossils every year. To date, the most primitive member of this privileged cohort is possibly the 2-meter-long (a little more than 6 feet) *Dilong paradoxus*, an animal whose anatomy reveals a distant kinship with the famous *Tyrannosaurus rex* of the Late Cretaceous of the United States and Canada. In addition to being much smaller than its 66-million-year-old North American cousin, the much older *Dilong* had relatively long arms and a hand with three fingers, unlike the short, two-fingered arms iconic of *Tyrannosaurus rex*. Among the handful of fossils of *Dilong*, one contains small portions of plumage on the

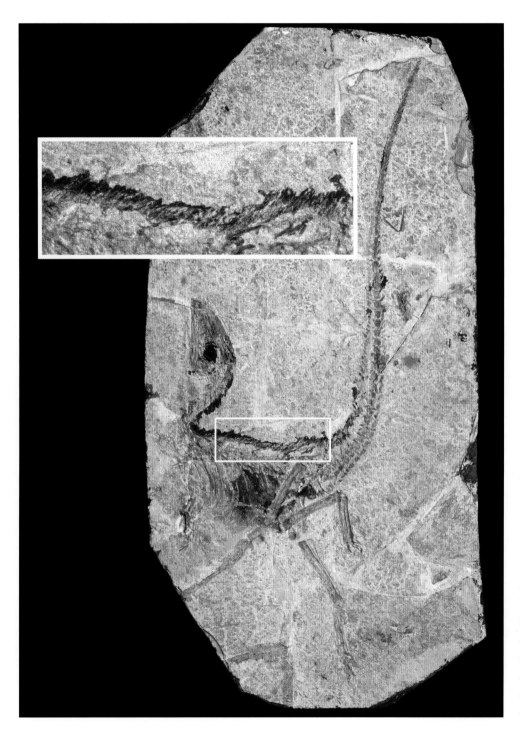

Unearthed in 1996, the first specimen of *Sinosauropteryx prima* showed that a dark halo of soft structures encircled much of the skeleton of this Jehol non-avian dinosaur. Details of their nature support the idea that these structures are protofeathers, early precursors to modern feathers. This 70-centimeter-long (almost 28 inches) fossil provided the first direct evidence that feathers evolved among non-avian dinosaurs. Since its discovery, more than 20 species of feathered non-avian dinosaurs have been found in China and elsewhere.

tail and the rear end of the lower jaw. Best preserved along the tail, these filament-like protofeathers are slightly shorter than 2 centimeters (less than an inch) and angled 30 to 45 degrees to the long axis of the bony tail; their appearance is in all respects similar to the protofeathers of *Sinosauropteryx*. The discovery of *Dilong* prompted the idea that even the mighty *Tyrannosaurus rex* could have been cloaked in feathers. Evidence of the external appearance of *Tyrannosaurus rex* is still wanting, but more recent fossil discoveries from the Jehol indicate that feathers were not restricted to small dinosaurs. *Yutyrannus huali* is another distant cousin of *Tyrannosaurus rex* but, unlike *Dilong*, *Yutyrannus* was close to 9 meters (30 feet) long and could have weighed about 1,400 kilograms (3,000 pounds). Long filament-like protofeathers, some measuring an impressive 20 centimeters (8 inches), are preserved in association with different parts of the skeleton of the few known specimens of this mighty dinosaur. While no specimen preserves a full set of protofeathers, these fossils suggest that most of the body of *Yutyrannus* was covered by structures similar to those found in *Dilong* and *Sinosauropteryx*.

While *Dilong*, *Yutyrannus*, and other relatives of *Tyrannosaurus rex* form a group that branched off the evolutionary stalk of theropods before the divergence of *Sinosauropteryx* and its kin, their overall plumage was in many ways the same. Also similar was the feathering found in a close, but larger, relative of *Sinosauropteryx*—the 2.5-meter-long (more than 8 feet) *Sinocalliopteryx gigas*. Size aside, these two dinosaurs are very similar to one another. The primitive, fluffy type of plumage of *Sinocalliopteryx* also covered much of its body and, like in *Sinosauropteryx*, it differed in length across the body. In both these animals, the longest of these simple, filament-like protofeathers cover the hip and the upper base of the tail; in *Sinocalliopteryx* they reach up to 10 centimeters (4 inches) in length. Both *Sinosauropteryx* and *Sinocalliopteryx* are close relatives of *Compsognathus*

Closeup of the filament-like protofeathers that covered the body of the fearsome *Yutyrannus hauli*, a 9-meter-long (almost 30 feet) relative of *Tyrannosaurus rex* from the Jehol Biota of China.

longipes, a much older dinosaur that lived next to *Archaeopteryx* some 150 million years ago in what is today's southern Germany. *Compsognathus* played a pivotal role in the early controversy about the origin of birds (and evolution in general) during the nineteenth century. Thomas Henry Huxley, the spirited proponent of the dinosaurian origin of birds, used the small *Compsognathus* to articulate some of his views about the ancestry of birds. While none of the fossils of this famous European dinosaur preserve any clear remnants of plumage, the discovery of its close relatives from the Jehol Biota supports the idea that it was also feathered. It is interesting to imagine the historical course that the winding ideas of bird origins would have taken had Darwin, Huxley, and their peers found themselves face to face with a feathered *Compsognathus.*

Evidence of plumage has also been found among herbivorous theropod dinosaurs. While theropods are commonly interpreted as predatory, and in fact many of them probably were, these dinosaurs show a wide range of feeding specializations. With small heads and typically toothless (except for some primitive forms that carried small teeth), the ornithomimosaurs are one group of theropods that has been regarded as herbivores. Recent discoveries have documented the presence of small filament-like protofeathers surrounding the skeletons of several specimens of an ornithomimosaur species—*Ornithomimus edmontonicus*—from 75-million-year-old rocks in Alberta, Canada. Another peculiar lineage of theropods, the therizinosaurs, also has consistently been interpreted as herbivores. These dinosaurs, with extended necks, small heads with leaf-shaped teeth, broad chests, long arms with powerful claws, and stout feet, ranged significantly in size. Two small species, *Beipiaosaurus inexpectus* and *Jianchangosaurus yixianensis,* are known from the Jehol Biota. At almost 2 meters in length (a little more than 6 feet long) these 125-million-year-old therizinosaurs were dwarfed by their gigantic cousins that lived millions of years later, toward the end of the Mesozoic Era; the largest of them, *Therizinosaurus cheloniformes* from Mongolia, had 3-meter-long (10 feet) arms and stretched more than 12 meters (40 feet) in length, its body weighing more than that of a male African elephant.

While we do not know the external appearance of this colossal dinosaur, the Jehol therizinosaurs give us a hint. Fossils of these modest-sized dinosaurs show that soft filaments similar in shape to those of *Sinosauropteryx* covered the body of *Beipiaosaurus.* Like in *Sinosauropteryx, Sinocalliopteryx, Dilong,* and *Yutyrannus,* these down-like structures possessed the branching arrangement and hollow base that are trademarks of avian feathers. On the forelimb of *Beipiaosaurus,* these downy protofeathers reached a length of up to 8 centimeters (more than 3 inches) and attached to the bones of the forearm in a tightly parallel pattern. The only known specimen of *Jianchangosaurus,* albeit preserving just a tiny portion of its plumage, suggests that a similar type of feathering occurred on this dinosaur as well. Combined, these fossils show that a coat of down-like feathers covered most of the surface of the bodies of therizinosaurs. Fossils of *Beipiaosaurus* also show that the external coat of these animals included a fewer number of longer (10 to 15 centimeters [4 to 6 inches]) and much broader, single filaments. These distinct structures, primarily distributed along the head, neck, and tail, have been interpreted as specialized protofeathers that played a role in the sexual displays of these dinosaurs.

Another branch of herbivorous to omnivorous theropods, the oviraptorosaurs, is well represented in the Jehol Biota. The anatomical characteristics of these animals, which include the renowned parrot-headed *Oviraptor* and its kin, indicate they were more closely related to birds than the therizinosaurs. What we know about their plumage

The filament-like appearance of the protofeathers attached to the forelimb of the therizinosaur *Beipiaosaurus inexpectus* and other early maniraptoran dinosaurs makes untenable the idea that feathers first evolved for the purpose of flight.

agrees with this interpretation. The best-known feathered oviraptorosaurs from the Jehol Biota, the turkey-sized *Caudipteryx zoui* and the somewhat larger *Caudipteryx dongi*, show that a minimum of 14 symmetrically large feathers projected from the outer half of their forelimbs. Most importantly, unlike their more primitive cousins— *Sinosauropteryx*, *Yutyrannus*, and *Beipiaosaurus*, among others—the forelimb feathers of *Caudipteryx* exhibit a central shaft and multiple thinner branches (called barbs) that project from either side of the shaft in a parallel arrangement, forming two large and flattened vanes. In these and other details, the feathers attached to the forelimbs of *Caudipteryx* have essentially the same appearance as a modern feather. Close examination also reveals that the outermost feathers of the forelimb were attached to the middle finger of the three-fingered hand of these dinosaurs, exactly the same finger that receives the attachment of the primary flight feathers in a modern avian wing.

In addition, these fossils show that the longest feathers, reaching up to 20 centimeters (8 inches), were flanked by shorter ones toward both the tip and the base of the forelimb, a pattern identical to what we see in the wings of present-day birds. The known fossils of *Caudipteryx* document that the down-like protofeathers typical of other feathered theropods also covered the bodies of the oviraptorosaurs. Yet, a curious tuft of long, vaned feathers fanned out from the end of the short tail of these dinosaurs. In fact, the bony tail of oviraptorosaurs—as well as those of therizinosaurs and ornithomimosaur theropods—ends in a short bone formed by the fusion of a few vertebrae (a small version of the avian pygostyle that ends the tail of modern birds). The long, vaned feathers attached to this bone and to other tail vertebrae in these dinosaurs might have offered some aerodynamic advantage, but given that the relatively short arms and overall proportions of these dinosaurs tell us that they were incapable of taking off, the elaborate feather fan at the end of the their tails might have played a much more important role in their sexual displays and during courtship.

Other Jehol relatives of *Caudipteryx* include the slightly smaller but somewhat stouter, *Protarchaeopteryx robusta* and the similarly sized *Similicaudipteryx yixianensis*. Whereas dozens of exquisitely preserved fossils of *Caudipteryx* have been unearthed from these Jehol sites, we know of only a few, incomplete specimens of *Protarchaeopteryx* and *Similicaudipteryx*. These fossils, however, add further insights that confirm the remarkable modern appearance of the plumage of oviraptorosaurs. A unique fossil of a young *Similicaudipteryx* has revealed that the tail feathers of this dinosaur molted through a "pin feather" stage in which the feather was enclosed in a protective sheath akin to the waxy-coated shaft of a modern, developing feather. These remarkable fossils have highlighted the great similarity between the vaned feathers of oviraptorosaurs, which seem to have evolved the basic molting mechanisms and structures that we see in modern feathers, and those of present-day birds.

The noticeable stiffness of the vaned feathers of oviraptorosaurs and their striking resemblance to those of living birds means that the Velcro-like system of interlocking hooked and grooved barbules (tiny branches projecting from the barbs), which maintain the cohesiveness of the vanes of modern feathers, had already evolved among these dinosaurs. These feathers also formed distinct airfoils at the ends of both the forelimbs and tail, which would have been capable of generating aerodynamic forces. Still, the forelimbs of these dinosaurs were proportionally too small to produce the lift necessary for taking to the air. Estimations of the wing loading (weight to wing area ratio) of these dinosaurs put this important aerodynamic parameter at around twice the value of the wing loading of *Archaeopteryx* and greater than that of any flying bird. The exquisite preservation of the plumage of *Caudipteryx* and its kin reveals a great deal of likeness with the feathers of living birds. Yet, the feathered forelimbs of this dinosaur were clearly unable to propel its heavy body into the air. It would take a substantial reduction in body size, coupled with an increase in the size of the feathered airfoils, to render the dinosaurian forerunners of birds airborne and ready to conquer the sky.

FLYING DINOSAURS The discovery of the first feathered dinosaurs from the Jehol Biota helped consolidate the idea that birds had their evolutionary origin from within the theropods, but later discoveries also showed that some of these Chinese dinosaurs were capable of flying. The best-known examples of these airborne dinosaurs belong to a group of small maniraptorans called paravians, which include the sickle-clawed dromaeosaurids, the lightly built troodontids, and the bizarre scansoriopterygids.

Abundant fossils of the dromaeosaurids *Microraptor zhaoianus* and *Sinornithosaurus millenii*, animals with sizes comparable to those of pheasants and turkeys, have been collected from Jehol rocks spanning in age from 125 to 120 million years ago. These fossils illustrate how the plumage of these dinosaurs evolved a degree of feather differentiation approaching what we see among living birds. In addition to the 2- to 5-centimeter-long (approximately 1 to 2 inches) downy protofeathers similar in design to those of other feathered dinosaurs, these Jehol dromaeosaurids show elongated feathers with distinct vanes that in some cases exhibit highly disproportionate widths. The most startling example of these asymmetric feathers is the spectacular *Microraptor*, which with a length of about 75 centimeters (30 inches), was somewhat smaller but much lighter than *Sinornithosaurus*.

If the bird-like appearance and reduction in size of *Microraptor* are without a doubt remarkable, the design of its plumage is absolutely stunning. Specimens of this dinosaur show that its tail carried a fan of elongated, vaned feathers, the terminal pair of which stretched to more than 22 centimeters (slightly more than 8 inches) in length. These notable fossils also indicate that *Microraptor* had long and slender wings formed by feathers with asymmetric vanes, a design that is often assumed to enhance aerodynamic competence. The attachment and distribution of flight feathers is in many ways similar to what we see in the wings of living birds: the feathers attached to the hand, the primaries, are longer but fewer than the secondaries, which are anchored across the forearm. Similarly, the farthest primaries are parallel to the direction of the fingers, while other flight feathers are angled with respect to the remaining bones of the forelimb. Yet, the most remarkable feature about *Microraptor*, and clearly the attribute that launched it to immediate stardom, is the presence of a set of vaned feathers—in some cases as long as 20 centimeters (8 inches)—projecting from its shin and foot.

Long, filament-like proto-feathers surrounded the body of the Jehol dromaeosaurid *Sinornithosaurus millenii.* Similar structures are found in many other non-avian theropod dinosaurs.

(*Opposite*)
The 75-centimeter-long (about 30 inches) *Microraptor zhaoianus* (*top*) displays sets of flight feathers attached to both the forelimbs and the hindlimbs, and a tuft of these feathers anchored to the end of its long bony tail. The 30-centimeter-long (12 inches) tail feathers of the much larger *Changyuraptor yangi* (*bottom*) are the longest known feathers of any non-avian dinosaur. Aerodynamic studies on these Jehol fossils indicate that flying non-avian dinosaurs used their feathered tails to enhance maneuverability during flight.

There is little doubt that the well-developed wings and tail of *Microraptor* afforded this animal sufficient lift and thrust to take to the air. Whether it was predominantly a glider that launched itself from an elevated perch, or an animal capable of flapping its wings and taking off from the ground, is not yet clear. Likewise, details of the functional significance of the astonishing hindwings have not been entirely worked out. The most reasonable interpretations reconstruct these feathers as trailing behind the lower leg and assisting in flight control. Another recent example of a feathered dromaeosaurid from the Jehol Biota, with hindwings like *Microraptor*, is the spectacular *Changyuraptor yangi*. At approximately 1.5 meters (5 feet) in length, this close relative of *Microraptor* was approximately 4 times heavier. While similar to its smaller cousin in many ways, *Changyuraptor* boasts the record of the longest feathers known for any dinosaur. The feathered tail of this dromaeosaurid formed an extensive flat surface in which the longest, central feathers reached approximately 30 centimeters (12 inches) long. Functional studies of *Changyuraptor* have demonstrated that hind-winged dromaeosaurids were able to maintain aerial proficiency even at sizes significantly larger than that of *Microraptor*.

In the past few years, much older fossils of small paravians with enlarged hindwings have also been unearthed from the Mesozoic of northeastern China. *Anchiornis huxleyi*

is the best known of these animals; its fossils have been collected from quarries dating to about 160 million years ago, in the Jurassic. While the definitive place of *Anchiornis* in the family tree of dinosaurs is not entirely clear, features of its skeleton suggest that it most likely represents a primitive troodontid, a paravian that is well known in the Cretaceous. Available fossils show that the forewings of *Anchiornis* were shorter and more rounded than those of the much younger *Microraptor*, and that, unlike in the latter, the feathers of *Anchiornis* do not seem to have asymmetrical vanes. Additionally, the hindwings of *Anchiornis* appear to form a smaller airfoil than those in *Microraptor*, and its tail lacks the long vaned feathers of the Jehol dromaeosaurids. Altogether, these features suggest that the quail-sized *Anchiornis* was possibly not as good a flier as the dromaeosaurids, which is not the same as saying that it may not have had a degree of aerodynamic competence. Feathered troodontids are also known from the Jehol Biota. These younger cousins of *Anchiornis* include the similarly sized *Jinfengopteryx elegans*. In the only known fossil of this animal, abundant protofeathers similar to those in *Sinosauropteryx* are preserved covering the neck, parts of the forelimbs, and the hips and thighs. Its bony tail, however, carries a set of long and symmetrically vaned feathers that become increasingly longer, forming a beautiful plumed tail. In contrast to its earlier cousin, *Jinfengopteryx* appears to lack hindwings, although it is difficult to say whether this is the result of the preservation of the only known fossil of this animal.

Perhaps the most peculiar of the feathered paravians from the Mesozoic of China are a poorly known group of tiny dinosaurs called scansoriopterygids, a name based on the scansorial, or climbing, capabilities these animals probably possessed. Until recently, the best known of these mysterious dinosaurs was *Epidexipteryx hui*, whose 25-centimeter-long (10 inches) skeleton was unearthed from Jurassic rocks comparable in age to those containing the slightly larger *Anchiornis*. Coated by filament-like protofeathers typical

The primitive paravian *Epidexipteryx hui* is one of the many feathered non-avian dinosaurs found in China during the past three decades. Soft feathers, preserved as a dark halo around its back, belly, and neck, surrounded the tiny skeleton of this bizarre, 160-million-year-old dinosaur and four long, tape-like feathers ornamented its tail.

of many other theropods, this bizarre little dinosaur is characterized by having a skull with few, large, and somewhat procumbent teeth, long forelimbs with unusually extended outermost fingers, and a relatively short tail that anchors four long, ribbon-like feathers.

More recently, the list of strange anatomical features of these dinosaurs has been augmented by the discovery of the stunning *Yi qi* from rocks coeval to those that yielded the skeleton of *Epidexipteryx*. *Yi qi* literally means "strange wing"; its forelimb preserves a rod-like bone projecting from the wrist and associated with the remnants of a "flight" skinfold. This discovery has prompted comparisons between scansoriopterygids and flying squirrels, bats, and pterosaurs, all of which bear similar slender bones extending from their wrists and supporting a membranous airfoil. The only known fossil of *Yi qi* also clarifies aspects of the plumage of the scansoriopterygids. It shows that these animals lacked vaned feathers, and that their bodies were covered instead by stiff filament-like structures formed by a relatively thick shaft, frayed into nearly parallel strands toward its terminal one-fourth; the ribbon-like feathers that ornamented the tail of *Epidexipteryx* (unknown for *Yi*) may well represent modified versions of the stiff, filament-like plumage of *Yi qi*. Collectively, the scansoriopterygids look a lot less bird-like than either the dromaeosaurids or the troodontids, but details of the anatomy of these curious dinosaurs continue to suggest that they represent a maniraptoran offshoot close to the origin of birds. *Yi qi*, with its unusual forelimb configuration, offers a glimpse at an aerodynamic body plan utterly different from those of birds and their winged, non-avian forerunners. Its discovery demonstrates the remarkable degree of evolutionary experimentation achieved by some of the lineages close to the origin of birds. Notwithstanding the aerodynamic idiosyncrasies of the scansoriopterygids, these little dinosaurs join a cohort of other feathered dinosaurs that highlight the enormous diversity of plumage and potential aerial strategies that evolved among the forerunners of birds.

The tiny little skeleton of *Yi qi*, bearing a coat of filament-like protofeathers, has surprised the paleontological community. The enlarged forelimbs of this bizarre maniraptoran exhibit a long, auxiliary wrist bone (*arrows*). Soft tissue associated with the forelimbs hints at the presence of a membranous airfoil supported (and presumably controlled) by this bony strut. This curious fossil suggests that scansoriopterygids might have evolved a singular mode of aerial locomotion among dinosaurs, gliding from elevated perches like some modern arboreal reptiles and mammals.

213

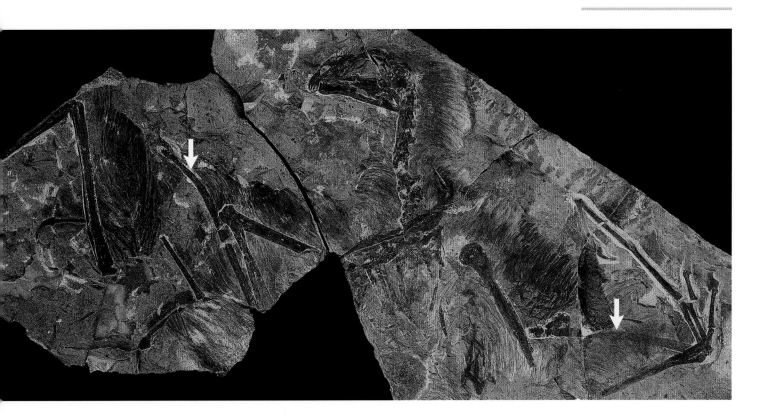

The astonishing diversity of feather structures known for different types of non-avian theropods indicates that true feathers, and/or fuzzy filaments considered as proto-feathers, covered the bodies of many of these dinosaurs. The distribution of the vaned feathers anchored to the limbs and tails of non-avian maniraptorans suggests that flight and tail feathers evolved first in the extremities of these dinosaurs and that these initial airfoils might have had the capacity to generate aerodynamic forces, although it is difficult to rule out the possible role of these feathers in courtship or other sexual displays. The repertoire of feathered airfoils, the limb and body proportions, and a number of details of the bones all suggest that some of the paravians closer to the origin of birds—*Microraptor, Changyuraptor, Anchiornis,* and perhaps the bizarre scansoriopterygids—were capable of some degree of aerial locomotion. As with many attributes long thought to be limited to birds, these discoveries have established the view that the flight of birds was also inherited from their small dinosaurian predecessors, which millions of years ago shared the sky with a diversity of ancient birds.

THE ORIGIN OF FEATHERS Feathers are the most complex skin structures of any vertebrate animal. Like hair, they are projections of the outermost layer of the skin, although made of a harder type of the protein keratin. Most feathers have a multi-branched architecture dominated by a long, central shaft that branches into numerous pairs of equally spaced barbs, the delicate projections that form the two vanes (or webs) of a typical feather. The tubular base of the shaft is embedded in the skin and anchored by a network of delicate muscles and ligaments, which control the movement of the feather. The barbs that branch from the shaft are in turn subdivided into tiny filament-like segments called barbules, which project outward on each side of the barb. In those feathers in which close-knit vanes form a compact surface, the barbules are further differentiated in such a way that those branching off one side of the barb carry microscopic hooks and those that branch off the opposite side are excavated by a groove. The Velcro-like latching of the hooked barbules of a barb onto the grooved barbules of its neighboring barb gives the vane its characteristic cohesiveness and its ability to generate the aerodynamic forces necessary for flight. In feathers without vanes—down feathers and other types of fluffy feathers—the barbules lack this hook-and-groove type of latching mechanism. Instead, the barbules of downy feathers form a mesh that is ideal for retaining air and providing insulation to the body.

Over millions of years, feathers have evolved into a multitude of types that are differentiated from one another by variations in the arrangement, size, and structure of the shaft, barbs, and barbules. Some types of feathers are even extinct. Having characterized certain groups of Mesozoic birds, these feathers are not present in their living counterparts. Most feathers can be classified as either contour or down. Contour feathers, including body, flight, and tail feathers, are the vaned feathers that cover most of the surface of the body. Flight and tail feathers are long and, in some instances, their vanes are asymmetric, with one clearly narrower than the other. Collectively, flight and tail feathers form the main aerodynamic surfaces that provide the lift and thrust forces necessary to propel birds through the air—they also help birds to maneuver and to control their movements during flight. Body feathers are contour feathers that are somewhat softer than either flight or tail feathers. While the top half of a body feather forms tightly

Down

Body

Flight

Rachis

Barb

Barbule

Most feathers of living birds are divided into flight, body, and down feathers. Flight feathers are characterized by having a shaft, or rachis, which branches into sets of barbs that in turn branch into sets of barbules. The cohesiveness of the feather vane is maintained by a microscopic system of hooks and grooves that lock the barbules of adjacent barbs in place. Body feathers are shafted, but they have this interlocking system only along their upper half. Down feathers lack a rachis and their barbs have barbules without hooks and grooves; they have a fuzzy appearance.

knitted vanes, its lower half is fuzzy and devoid of interlocking barbules. Body feathers are ideally designed for waterproofing the bird's body and keeping it warm.

Unlike contour feathers, downy feathers are completely fluffy, are usually smaller, and have fewer barbs. The loosely arranged, disconnected barbules of downy feathers form a mesh that traps pockets of air next to the skin, which provides insulation against heat loss. Down feathers thus work cooperatively with the body feathers, sheltering the bird from wind, sun, rain, and the surrounding environment. A variety of even further specialized feathers—bristles and filoplumes, among others—perform more specific functions such as sensory acuity, signaling, and foraging.

Traditionally, interpretations of the evolutionary origin of feathers were largely governed by the now refuted notion that feathers derived from the scales of ancestral reptiles that became longer, deeply fringed, and more flexible over millions of years. In the past two decades, however, studies on modern feathers and scales have severely undermined this view by recognizing important differences in the development of these skin structures. During the same time, many fossil discoveries have documented the presence of feather-like structures that are unlike any known scale in the dinosaur precursors of birds. The combination of these findings has established that modern feathers

Flight feathers of modern appearance form the wing of this 131-million-year-old enantiornithine bird from the Jehol Biota of northeastern China. A diversity of modern-looking feathers and at least two types of extinct feathers have been discovered in association with the skeletons of fossil birds from these Chinese sites.

evolved as a series of intricate structures originating sequentially during the long evolution of dinosaurs (and perhaps even before) and that most of these evolutionary novelties share neither a molecular nor a developmental resemblance to reptilian scales. Nonetheless, the question of whether the filament-like protofeathers that cover the bodies of many Mesozoic theropod dinosaurs have a much more ancient origin remains controversial.

As mentioned earlier, superficially similar structures have been found in association with the skeletons of small ornithischian dinosaurs such as psittacosaurs and hetero-dontosaurs. In these dinosaurs, numerous long filament-like structures project upward from the back and the tail. Particularly intriguing is the ornithischian dinosaur *Kulin-dadromeus zabaikalicus* from the Middle to Late Jurassic of Russia, in which three types of skin structures have been recognized. While these structures do not resemble the vaned feathers of modern birds, some have a superficial similarity to the filament-like protofeathers of *Sinosauropteryx* and various Mesozoic theropod dinosaurs. These and other discoveries have prompted the claim that early precursors of modern feathers evolved among the first dinosaurs or even before them.

Such an assertion, however, has encountered resistance in a recent comprehensive study of the different groups of dinosaurs in which remnants of their external appearance are preserved; this revision has demonstrated that the filament-like skin structures of ornithischians such as *Kulindadromeus*, psittacosaurs, and heterodontosaurs are more the exception than the norm and, hence, that the earliest dinosaurs and their predecessors were most likely scaly. Nonetheless, the picture remains complex, especially when considering that skin structures also believed by some to be feather precursors have been found covering the head and body of several fuzzy-coated pterosaurs, the flying reptiles of the Mesozoic Era. Detailed comparisons between these skin structures and those found among dinosaurs are difficult, and there is no consensus about whether they are the same type of structure or not (many pterosaur experts use the term "pycnofiber" to refer to these structures and to emphasize their interpretation as neither hair nor feathers). While all these intriguing fossils pose the question of whether early precursors of modern feathers first evolved during the Triassic diversification of archosaurs, at the dawn of dinosaurs or even earlier, the information available to untangle such a conundrum remains quite limited.

While new discoveries and studies of the past two decades have greatly improved our understanding of the evolutionary beginning of feathers, these advances have fallen short of solving the debate surrounding the role played by these unique structures when they first evolved many millions of years ago. For more than a century, scientists attempting to explain their functions have elaborated an array of ideas that range from those focusing on aerodynamic considerations—feathers originating for the purpose of flying—to others emphasizing display, insulation, waterproofing, foraging, and other functions known to occur among birds living today. But the primal function of feathers is still a big open question, although some alternative explanations have been dismissed. One important outcome from the discovery of a range of feathered non-avian dinosaurs from the Jehol Biota and elsewhere is the realization that feathers did not evolve in the context of flight. This conclusion is well supported by the fact that the most primitive of these dinosaurs have forelimbs that are proportionally much shorter than those of flying birds and filament-like protofeathers that lack the vanes necessary to generate aerodynamic forces. Therefore, fossils of the most primitive feathered dinosaurs clearly indicate that at their onset feathers must have played a role other than flight.

Today, most researchers agree that the original purpose of feathers was either insulation or display. The former idea argues that feathers evolved as a solution to reduce heat loss in animals that were developing higher metabolic rates and on the road to becoming warm-blooded. Several lines of evidence support the notion that most dinosaurs had more elevated body temperatures and faster growth rates (characteristic of

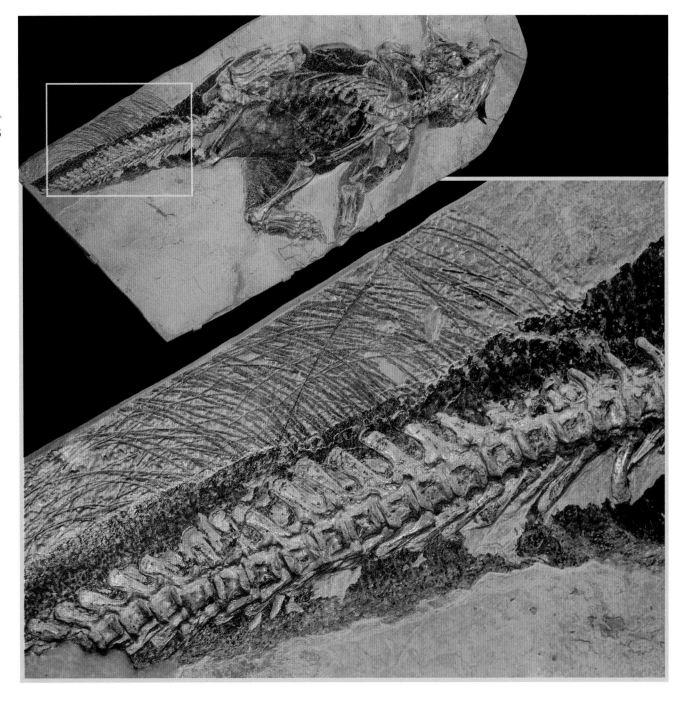

Filament-like skin structures projecting from the tail of ornithischian dinosaurs such as the psittacosaurs have prompted claims that protofeathers might have originated at the dawn of the dinosaurs or even before the origin of these animals.

warm-blooded animals) than those typical of modern reptiles. Geochemical analyses of the tooth enamel and the eggshells of non-avian dinosaurs, as well as studies of the microstructure of their fossilized bone tissue, indicate that the internal body temperature and rates of skeletal growth of these animals were more similar to those of living birds and mammals, which are warm-blooded. For example, a recent study of chemical isotopes in the eggshell of an oviraptorid maniraptoran estimated the internal body temperatures of these dinosaurs to be about 32 to 35 degrees Celsius, above the typical 27 to 28 degrees Celsius of crocodiles and below the 38 to 43 degrees Celsius of present-day birds. Other recent analyses have argued that additional physiological advances

Shafted feathers, like these from the wing of the Jehol bird *Confuciusornis sanctus*, derived from more simple feathers of non-avian theropod dinosaurs. The fact that the feathered predecessors of birds were obligatory ground-dwellers indicates that the original function of feathers was not flight.

toward the development of warm-bloodedness were taken in conjunction with the evolution of vaned feathers and closer to the origin of avian flight. A diverse suite of inferences has recently shown that the metabolic rates of non-avian dinosaurs, while possibly varying from group to group, were intermediate between living cold-blooded and warm-blooded animals. In other words, non-avian dinosaurs might have evolved a metabolic physiology that was closer to that of their living relatives but without necessarily reaching the warm-bloodedness typical of modern-day birds. Therefore, the fact that the plumage of the most primitive feathered dinosaurs is composed of simple, filament-like protofeathers is fully consistent with the idea that these structures first served purposes related to insulation, very much in the same way hair and down insulate the body of warm-blooded mammals and birds, respectively.

Alternatively, scientists who lean toward explaining the primary function of feathers within the context of behavioral display argue that these structures evolved to facilitate identification of suitable mates or to boost competition for the fittest mates. Mesozoic fossil feathers with gray and brown monochromatic arrangements—light and dark dots along barbs, lighter and darker stripes over an entire feather, or throughout the tail, as in the Chinese feathered theropod *Sinosauropteryx*—have been known for years. These fossils document the antiquity of color schemes in feathers but not until recently have we started to get a better idea of the specific colors that adorned the bodies of the earliest feathered dinosaurs. One line of evidence rests on microscopic studies that have been capable of identifying the shape and other characteristics of the cellular storage for melanin, the primary pigment controlling the black, brown, and reddish tones of feathers. In living animals, different types of melanin, each responsible for a specific color, are stored in cellular organelles called melanosomes. The shape of these microscopic capsules varies according to the type of melanin they carry: long with rounded ends for the black and brown varieties, and more circular for the reddish type. Recent microscopic analyses of the feathers of non-avian dinosaurs, including the 160-million-year-old *Anchiornis*, have identified these types of melanosomes and, hence, mapped the approximate color of the plumage of these animals.

More comprehensive studies involving a wide range of vertebrates—lizards, turtles, crocodiles, mammals, and birds—have shown that melanosome shape also varies from one animal to the other, and among different types of skin structures (scales, feathers, hair, and others). These studies have recognized an important difference in the diversity of melanosome shape between lizards, turtles, and crocodiles, on the one hand, and birds and mammals, on the other hand. Collectively, the melanosomes of birds and mammals differ in shape much more than those in lizards, turtles, and crocodiles. In dinosaurs, this shift in the diversity of the structures interpreted as melanosomes becomes visible between those non-avian dinosaurs with simple, filament-like proto-feathers (*Sinosauropteryx*, among others) and those that carry vaned feathers. The greater diversity of melanosome shape among non-avian maniraptorans with vaned feathers, and their avian descendants, suggests that an increase in melanin-based coloration took place prior to the origin of birds.

The identification of melanosomes in ancient feathers has not been free of controversy. Detractors argue that the recognition of melanosomes in the fossil record needs to be supported by geochemical analyses because the general shape and size of the presumed structures overlaps with that of bacteria, which are intimately connected to the decay process and thus, they are expected to be present in the fossilized soft tissues of dinosaurs. These researchers emphasize the need to validate claims for the presence of fossilized melanosomes through chemical fingerprinting, namely, through the detection of the molecular characteristics of the different types of melanin.

Other recent approaches have also shed light on the colorful appearance of the dinosaurs that were closer to the origin of birds and their avian relatives, albeit again, not without controversy. X-ray techniques using sophisticated instruments such as cyclic particle accelerators, the same type of instrumentation used to study subatomic particles, have relied on the detection of residues from copper and other trace metals preserved in the feathers of non-avian dinosaurs as indicative of melanin concentration. Like the melanosome investigations discussed above, trace metal studies provide maps of the distribution of melanin, particularly the black and brown types, in the plumage of non-avian dinosaurs and ancient birds. While tantalizing, these studies are also susceptible to the same types of validation problems that plague those involving the fossilized structures presumed to be melanosomes; consequently, one should be careful before attributing too much confidence to the colorful reconstructions that have been presented for a variety of the feathered dinosaurs from the Jehol Biota.

Controversy notwithstanding, all these studies suggest that in addition to the observed variation of feather shapes, different patterns of color might have brightened the plumage of many feathered dinosaurs. Furthermore, inferences based on the genealogic placement of non-avian dinosaurs with respect to living birds, crocodiles, and mammals suggest that the forerunners of birds might have been able to see the wide spectrum of colors that present-day birds do. If early maniraptorans and other non-avian dinosaurs were able to see red, blue, green, and even UV portions of the light spectrum, feathers might have been far more colorful than what we may think based on the studies of color proxies. The discovery of such proxies for true color and the notion that non-avian dinosaur feathers might have been just as colorful as those of their living descendants have stimulated the idea that feathers could have evolved within the context of display, particularly when considering that feather coloration plays a paramount role in courtship and other display behaviors of modern birds.

The avian feather is a remarkable structure, a wonder of nature originating as the result of many millions of years of evolutionary experimentation by the different lineages of animals it adorned. The origin of this intricate structure remains obscure, hampered by the difficulties of interpreting the true nature of fossilized structures associated with the skeletons of many different types of extinct animals. Likewise, the primary role played by the earliest feathers remains controversial. Interpretations of the primal function of complex evolutionary novelties are always challenging and, in the case of feathers, we cannot rule out the possibility that these structures might have had more than one original purpose. Despite these shortcomings, the fossil record speaks unmistakably about how feathers of different colors and different shapes ornamented the bodies of many animals that for years have been classified within familiar groups of non-avian dinosaurs.

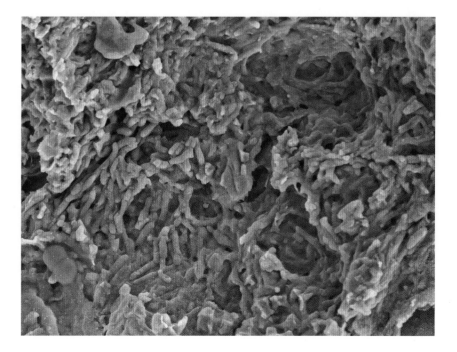

The recent discovery of melanosomes, microscopic capsules carrying melanin pigments, preserved in the feathers of a variety of non-avian dinosaurs and early birds has given us the first glimpse at the color of the plumage of these animals, albeit not without controversy. Photograph by Dr. Jeremy Cook from a specimen prepared in the laboratory of Prof Ruth Bellairs.

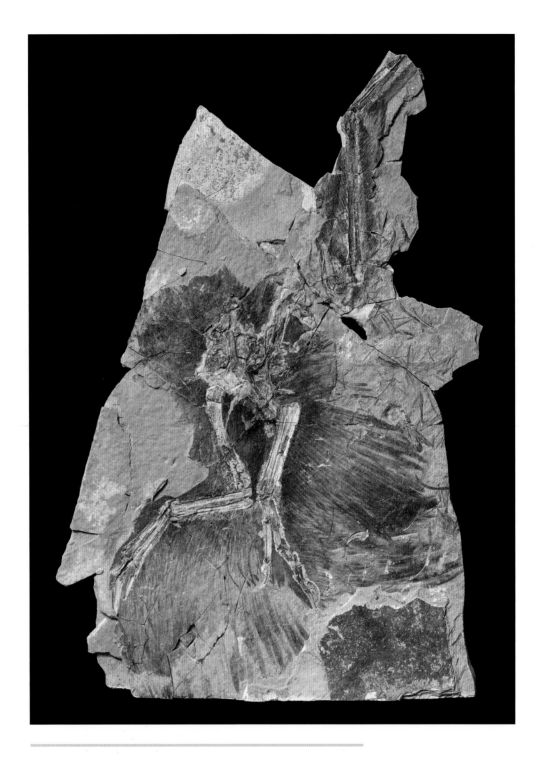

Mapping the diversity and concentration of melanosomes preserved in the feathers of *Anchiornis huxleyi* provides a picture of what the plumage of this Jurassic maniraptoran dinosaur might have looked like: a dark gray body, with a reddish crown and small rusty spots on its face, and whitish flight feathers with black spangles. Not surprisingly, these interpretations have been contentious. Further studies and additional specimens are needed to confirm the accuracy of depictions such as this one.

THE MESOZOIC AVIARY Despite the celebrated discoveries of the Late Jurassic *Archaeopteryx lithographica* and other Mesozoic-aged birds in the nineteenth century, the fossil record of early birds remained scant for the following 100 years. For much of the twentieth century, such limited evidence led to the idea that the diversity of birds during the Mesozoic was small and constrained by the evolutionary success of their dinosaurian predecessors. Nonetheless, paleontological discoveries of the past three decades, in China and elsewhere, have drastically changed our earlier views about the diversity and evolutionary patterns of early birds. Among other things, these findings have revealed that an enormous variety of these animals actually inhabited the same environments that were home to many different kinds of non-avian dinosaurs. These discoveries, however, have not yet knocked *Archaeopteryx* off its perch; fossils of *Archaeopteryx* are still regarded as the oldest known evidence of birds, although such a prominent status may be lost at any moment—prolific excavations of the Middle to Late Jurassic Daohugou Biota of northeastern China might someday discover an even older bird and with it expand the roster of the birds that lived next to the large dinosaurs of the Mesozoic Era.

The story of *Archaeopteryx* began with a single feather, trapped in stone 150 million years ago and unearthed in 1861 at a limestone quarry near the village of Solnhofen, in southern Germany. This pristine, ancient fossil displayed all the trademarks of an avian flight feather. During the same year, a partial skeleton cloaked in long-vaned feathers was collected from the same quarry. Among its bones was a boomerang-shaped wishbone, somewhat different from those of its modern relatives but undoubtedly avian. Despite the presence of feathers and an avian-looking wishbone, these fossils were not universally accepted as indicative of the very ancient pedigree of birds. The reptilian features of the skeleton puzzled some scientists, who argued that the fossilized remains of this animal could well belong to a reptile with bird-like feathers. Yet, whether it was considered a bird-like reptile or a reptile-like bird, *Archaeopteryx* was quick to enter the heated controversy surrounding the new ideas about the origin of species and Charles Darwin's views of common descent: the few skeletal features that set this ancient animal apart from the theropod dinosaurs known at the time engulfed its fossils in the nineteenth-century debate about the rise of birds.

More than 150 years after it was first discovered, the remarkable fossils of *Archaeopteryx* still remain the cornerstone for much of our modern understanding of the evolutionary origin of birds. These rare fossils—thirteen specimens (including the isolated feather found in 1861)—have provided a clear picture of the anatomy of the earliest birds and their intermediate appearance between their theropod forerunners and their modern counterparts. Very similar in skeletal design to one another but showing significant differences in size and developmental age, most specimens of *Archaeopteryx* have at some point in history been classified as distinct, but very closely related species. While there is some anatomical evidence for what could be considered a different species, the degree by which these skeletal differences are related to the various developmental stages (and sizes) of the specimens is not well known. In the end, all the fossils of *Archaeopteryx* may well belong to a single species, *Archaeopteryx lithographica*.

At about 50 centimeters (20 inches) in length, the dimensions of a midsize specimen, *Archaeopteryx* sported conical teeth on its jaws, a long and sharply clawed three-fingered hand, and an elongate bony tail that accounted for half the length of the animal. Despite the presence of large wings formed by powerful flight feathers, the skeleton of *Archaeopteryx* lacked important modifications usually associated with powered

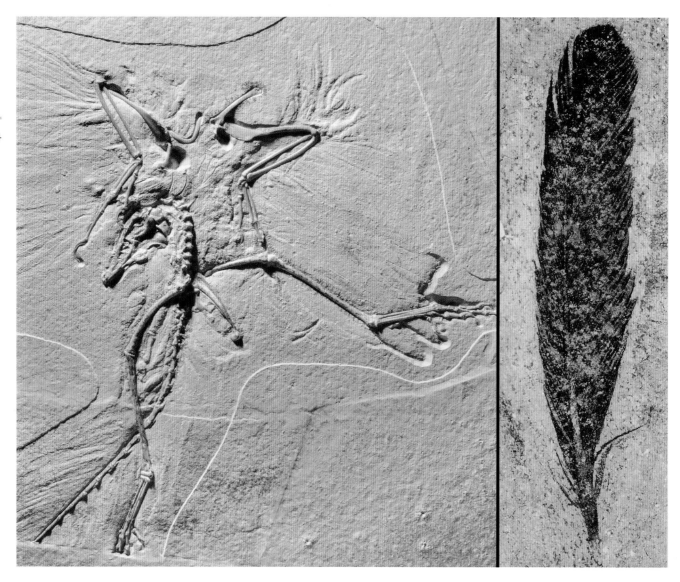

The plumage of the 150-million-year-old *Archaeopteryx litho-graphica* was in many ways identical to that of its modern counterparts. Studies of the preserved melanosomes of its plumage suggest it was primarily black in color.

flight: it had a long trunk, a short and weak sacral region, and no evidence of a breast-bone in all known fossils, thus indicating that the muscles that powered its wings did not have a strong anchor. Additionally, X-ray computed tomography (CAT) scans of its cranial cavity reveal that the size of its brain was proportionally much smaller than those of its living counterparts.

Studies of the past couple of decades have continued to emphasize the "missing link" nature of *Archaeopteryx* that made it an icon of evolution the very same year it was first discovered. Our current understanding of this animal, however, tells us that in many ways it was very similar to its non-avian maniraptoran relatives. Characteristics of its bone tissue indicate that, like its forerunners, *Archaeopteryx* grew over multiple years and that it reached reproductive maturity before becoming fully grown. Other analyses have found that the relative size of its brain was significantly bigger than that of typical non-avian theropods but comparable to the relative dimensions of the brain in some maniraptoran groups (oviraptorids, troodontids, and others). *Archaeopteryx* and its man-iraptoran predecessors also show increased complexity in brain architecture, with a suite of features indicating that these animals were neurologically ready for the demands of flight. In fact, anatomical details of the brain coupled with the modern-looking plumage and large wings of *Archaeopteryx* all support the notion that this animal (as well as some of its close maniraptoran relatives) might have been able to fly. How well it flew, however, and whether it was able to take off from a standstill, from an elevated perch, or after a takeoff run, are questions that still remain controversial. While recent advances continue to demonstrate a remarkable likeness between *Archaeopteryx* and its most immediate maniraptoran relatives, they have also cemented the notion that this ancient animal is a step closer to modern birds than any of its dinosaurian forerunners. More than 150 years after it was first unearthed, *Archaeopteryx* remains an enduring symbol of the evolutionary transition between early dinosaurs and present-day birds. Like an avian version of the Roman god Janus, one half of this renowned fossil bird looks backward while the other half looks forward.

Much younger birds, but still with primitive long bony tails and rudimentary flight-related skeletal features, are known from the Jehol Biota. These birds, 25 to 30 million years younger than *Archaeopteryx*, include *Jeholornis prima* and other turkey-sized fossils that are classified as jeholornithids. A suite of features from their skeletons indicates that these animals are evolutionarily closer to modern birds than *Archaeopteryx*. Their long bony tails, however, show that such a reptilian holdover, with the role it played in both terrestrial and aerial locomotion, persisted for many millions of years after the origin of birds. Some have argued that the jeholornithids had arboreal lifestyles, foraging and spending most of their time in the canopy of the Jehol forests. Nonetheless, such interpretations are at odds with the anatomy of the feet of these birds, which do not reveal any perching specialization.

Similar in size and sometimes regarded as a member of the early cohort of long bony-tailed birds is *Rahonavis ostromi*, the remains of which are found in rocks formed about 70 million years ago on the island of Madagascar. Fossils from this animal show that it had long wings with bones that anchored stiff flight feathers and powerful

The skeleton of the Jehol bird *Jeholornis prima* shows a degree of modernization with respect to its older and more primitive relative, *Archaeopteryx lithographica*. The bony tail of *Jeholornis* was longer.

THE EARLY EVOLUTION OF BIRDS

raptorial feet. Like the African goshawk and other predatory birds of today, the Late Cretaceous *Rahonavis* had an enlarged recurved claw on its second toe, presumably used to efficiently kill its prey. However, the identification of *Rahonavis* as a bird has been controversial; some researchers argue that this animal should be classified as a dromaeosaurid theropod, belonging to a group of long-snouted dinosaurs primarily known from the end of the Cretaceous in South America. Additional studies and discoveries may be needed to settle this issue and to determine whether a lineage of birds with long bony tails persisted for millions of years into the end of the Cretaceous.

The fact that the fossils of *Rahonavis* have been alternatively considered as either birds or non-avian maniraptorans highlights once again the extensive evolutionary experimentation that took place during the transition from non-avian theropods to modern birds. The fossil record tells us that virtually every lineage involved in this evolutionary transition achieved some degree of birdness, whether it is manifested in the skeleton, the plumage, the physiology, or the behavior of these ancient animals. This phenomenon helps to explain how many of these transitional fossils have been repeatedly placed on either side of the increasingly blurred divide separating birds from their dinosaurian predecessors. Much remains to be learned about the different kinds of birds with long skeletal tails that branched out following the evolutionary divergence of *Archaeopteryx*. Yet, the available fossils hint at an undiscovered diversity of these primitive forms that preceded the evolution of birds with an abbreviated tail composed of fewer vertebrae and ending in a bony stump called the pygostyle (the structure that supports the "pope's nose" and the tail feathers attached to it).

The abbreviation of the skeletal tail was a paramount transformation during the early evolution of birds. Such an evolutionary milestone had profound consequences for the mechanics of how these animals ran, stood up, and flew. Yet, the details of this evolutionary transition are far from clear, although it most likely involved important changes in the expression of regulatory genes acting on the regionalization of the bony tail. Twenty million years after *Archaeopteryx*, the fossil record shows the abrupt appearance of a diversity of birds carrying a pygostyle at the end of a shortened bony tail. Critical transformations in skeletal and muscular architecture were coupled with innovations in plumage and a substantial reduction in size, perhaps a consequence of important changes in growth rates and development. Future discoveries of fossil birds from either the end of the Jurassic or early after the beginning of the Cretaceous are likely to clarify the sequence of evolutionary steps that led to the origin of the reduced bony tail typical of most birds, but today this critical transition remains clouded in mystery.

One fossil that has provided some evidence, albeit not without controversy, is the 125-million-year-old *Zhongornis haoae*. The relatively short tail of the skeleton of this small bird is composed of 13 short vertebrae (many less than in either *Archaeopteryx* or *Jeholornis*), and it lacked a pygostyle. In modern birds, their abbreviated tail is typically composed of 5 to 8 individualized vertebrae that are followed by a pygostyle formed by the early-age fusion of typically 3 to 6 vertebrae. Thus, the number of vertebrae in the tail of *Zhongornis* agrees with the upper range of vertebrae that make up the tail of living birds. Studies focusing on the embryonic development of chickens and other birds show that during early growth, the tail of present-day birds becomes shorter in relation to other parts of the skeleton. The anatomy of the tail of *Zhongornis* and our understanding of how the bony tail of modern birds develops during their embryonic phases suggest that during the early evolution of birds, their bony tail became abbreviated prior to the origin of a pygostyle for support of tail feathers.

As pointed out above, the interpretation of *Zhongornis* as intermediate between long-tailed birds and those with a short bony tail ending in a pygostyle does not lack its detractors. A recent study has claimed that *Zhongornis* is in fact a scansoriopterygid, the peculiar lineage of non-avian maniraptorans that includes the small *Epidexipteryx hui* and *Yi qi* from the Jurassic of northeastern China. While the latter interpretation is by no means uncontroversial, it makes us wonder whether *Zhongornis* could not be another example of the evolutionary experimentation toward birdness that is so manifest during the transition from non-avian theropods to birds. Once again, in regard to the origin of the short bony tail of birds, we need more discoveries to cut through what after decades of unearthing spectacular fossils still remains an evolutionary Gordian knot.

The oldest records of birds with a pygostyle are exclusively known from the Jehol Biota. These fossils show that as early as 131 million years ago, several lineages of birds had already evolved the abbreviated structure of the tail that characterizes all of their later descendants. The wide array of designs of the skull, teeth, wings, and feet of the

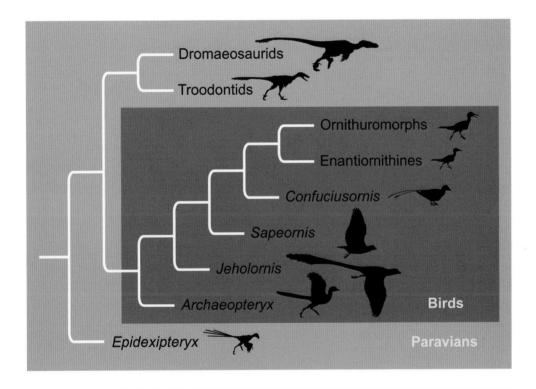

The evolutionary relationships among early birds and their place within the family tree of dinosaurs have been greatly clarified by the numerous discoveries of the Jehol Biota. The Jehol fossils have filled an enormous gap that existed between *Archaeopteryx lithographica* and birds much closer to the origin of modern birds. The Jehol Biota includes spectacular fossils of long bony-tailed jeholornithids (*Jeholornis prima* and kin); the primitive short bony-tailed *Sapeornis chaoyangensis* and *Confuciusornis sanctus*; a wealth of enantiornithines; and many primitive members of the ornithuromorphs, the group that includes all modern birds.

ancient Jehol birds, coupled with exceptional discoveries documenting what they ate, indicate that even at this early phase of their evolutionary history, birds had specialized into a variety of ecological niches. At the same time, a host of novel features in the wings, shoulders, and tail suggests that soon after *Archaeopteryx*, some birds evolved flying abilities not that different from the ones that amaze us today. The superior aerodynamic capabilities of these birds must have unlocked a wide range of ecological and evolutionary opportunities that are mirrored in the multiplicity of lifestyles and feeding strategies inferred from their fossils.

The Jehol rocks also yield abundant evidence of the most primitive groups of birds with an abbreviated tail ending in a pygostyle: the confuciusornithids and the sapeornithids. When compared to one another, these large birds—the crow-sized *Confuciusornis sanctus* and the even larger *Sapeornis chaoyangensis*—show varying degrees of primitiveness. *Confuciusornis sanctus* and other confuciusornithids had a skull with a powerful, toothless beak covered by a keratinous sheath. Most other Mesozoic lineages of

Laboratory experiments have revealed that the genes responsible for tooth formation still remain largely intact–albeit dormant–in the modern avian genome and that tooth growth can be activated or blocked on specific tooth-bearing bones by tinkering with a few regulatory genes. These investigations help us understand the genetic basis that determined the independent loss of teeth on different jawbones of various Mesozoic avian lineages. DNA studies have also dated the evolutionary event that suppressed the development of teeth among living birds to about 115 million years. Fossils of *Confuciusornis sanctus*, like the one shown here, illustrate the most archaic of these evolutionary events in birds.

birds are toothed, even if the fossil record shows teeth giving way to a completely keratinous beak (with the complete loss of dentition) in at least five separate lineages of extinct Cretaceous birds, the confuciusornithids representing just one and the oldest of them. Despite having a beaked and superficially modern jaw, the confuciusornithid skull had a very primitive architecture, which in many ways resembled that of its dinosaurian predecessors. The forelimbs of these birds were also very primitive, with three clawed fingers and other holdovers from earlier birds and their dinosaurian forerunners.

In contrast to the confuciusornithids, the upper jaws of sapeornithids were toothed, their teeth having distinct spade-shaped crowns, but their lower jaws were toothless. *Sapeornis chaoyangensis* and kin, however, show a more advanced forelimb than that of the confuciusornithids but a shoulder of more primitive appearance. Furthermore, unlike the confuciusornithids, which have well-formed breastbones, no fossils of sapeornithids have been found with a breastbone, which suggests that, like in *Archaeopteryx*, this element remained cartilaginous throughout the lives of these birds (some researchers have even claimed that these birds lacked a sternum altogether). Arguments about which of these two lineages—confuciusornithids or sapeornithids—is more primitive, and thus more removed from the ancestry of present-day birds, are split. Yet, collectively, their fossilized remains give us a clear picture of the basic appearance of the most primitive short bony-tail birds.

Little is known about the lifestyles of these early short bony-tailed birds. The proportions between the bones of the middle toe, a ratio that is sensitive to the different lifestyles of present-day birds, are comparable to those of birds that spend as much time foraging on the ground as in trees. Unfortunately, despite the hundreds of fossils of *Confuciusornis sanctus*, not a single one has given us reliable evidence of its diet. One study argued that it ate fish because one fossil was found overlaying a small fish, but such associations are common in the Jehol Biota and there is no evidence that this particular specimen of *Confuciusornis* held fish remains inside its gut. In contrast, structures interpreted as either fruits or seeds have been found in both the gut and the region of the crop of many fossils of *Sapeornis chaoyangensis*. Such findings provide direct evidence indicating that this early short-tailed bird ate either fruits or seeds, and most likely both. These discoveries are consistent with the occurrence of clusters of stomach stones within the belly of some *Sapeornis* specimens, which are common among living birds that eat grains, fruits, and other parts of plants. The stout beak of *Confuciusornis* and its kin also suggests that these birds could have used their powerful jaws to crack fruits, grains, or other hard plant material. Nonetheless, unlike the case of *Sapeornis*, no confuciusornithid specimen has ever been found with either the remains of food in its gut or with stomach stones. What was on the menu for this most abundant Mesozoic bird still remains elusive.

The Early Cretaceous Jehol Biota also documents the earliest representatives of another ancient group, the enantiornithines, which share a closer kinship with living birds than other primitive short bony-tailed birds. Fossils of these birds have also been collected from Early Cretaceous sites in Spain but from rocks that are slightly younger than those containing their oldest records in China. First recognized from Argentine fossils discovered in the 1970s, enantiornithine fossils have now been unearthed from every continent except Antarctica. Represented by more than 70 named species, and generally abundant in most sites that contain Cretaceous birds, the enantiornithines constitute the most diverse group of pre-modern birds. Remains of these birds are found

Confuciusornithids (*Confuciusornis sanctus*) and sapeornithids (*Sapeornis chaoyangensis*), both groups of early birds well represented in the Jehol Biota, are the most primitive known short-tailed birds. Their fossilized remains support the interpretation of these animals as diurnal seed- or fruit-eaters that primarily foraged on the ground.

Sapeornis chaoyangensis

Confuciusornis sanctus

from about 131 million years ago to the end of the Mesozoic, some 65 million years later. Displaying a remarkable range of sizes—as tiny as the smallest songbirds or as large as vultures—these birds are known to have lived in a variety of habitats, including deserts, forests, marine coastlines, and islands.

Like most birds that lived during the Mesozoic, the majority of enantiornithines would have looked very unfamiliar to the typical birder. Most of them had toothed jaws and other features pointing to a muscular skull with a relatively strong biting force. The forelimbs ended in two clawed fingers, the tips of which most likely emerged from the plumage. Yet, their skeletons show a series of key transformations revealing a significant step toward the anatomy of today's birds. Some of these new features include the shortening of the hand and fingers as well as changes in the proportions of the wing bones and the design of the shoulder. The architecture of the enantiornithine shoulder and the size and characteristics of their breastbones tell us that powerful flight muscles controlled the wings of these birds. The diversity of wing proportions and sizes found among them further suggests that their flight capabilities varied substantially, as did the flight modes they evolved during their multimillion-year history.

Not surprisingly, enantiornithines also evolved important aerodynamic innovations in their plumage. The exceptional preservation of many of their fossils, from China, Spain, and elsewhere, reveals that the primary and secondary feathers of the wing had sizes and numbers matching those of present-day birds. Soft tissue preservation points to great similarities between the muscles and ligaments controlling the movements of flight feathers in enantiornithines and their modern counterparts. In addition, the enantiornithine wing evolved a key aerodynamic structure called the alula, a small tuft of feathers attached to the thumb, which helps birds generate the additional lift needed during takeoff and landing. The significant transformations of the skeleton and plumage of these birds suggest that even at the onset of their ancient evolutionary history,

Spectacular fossils of early birds slightly younger than those from the Jehol Biota have been found in north-central Spain. Like their Jehol counterparts, many of these birds are small enantiornithines, the remains of which are exquisitely preserved. This aggregate of four tiny skeletons, all belonging to juvenile enantiornithines, has been interpreted as a pellet.

enantiornithines were able to take off from a standstill position and maneuver in ways similar to those seen among living birds. It is most likely that the evolution of these enhanced flying capabilities played a key role in the evolutionary success of the enantiornithines, which by about 120 million years ago seem to have risen to dominance.

The rich fossil record of enantiornithines has also given us a better understanding of their lifestyles. The feet of many of these archaic birds show specializations indicative of perching capabilities, which suggests that most of them lived in trees, like many of their living analogues. Gut contents preserved in some of these fossils—freshwater invertebrates and droplets of sap—give us a rare glimpse of their varied diet, although the vast majority of their fossils provide no direct evidence of their feeding behaviors. Most inferences about what these birds ate are based on the anatomy of their skulls and teeth. These interpretations suggest that while certain enantiornithines fed on grains, others gobbled insects and a few probed soft mud in search of worms and other small invertebrates.

Few enantiornithine eggs and embryos are known to date, and they are difficult to identify as belonging to any of the known species based on adult specimens. A recent study of egg shape among birds, non-avian dinosaurs, and other reptiles has revealed that the eggs of the pre-modern birds that lived during the Mesozoic, enantiornithines included, were proportionally longer than those of living birds and more similar to those of their non-avian theropod predecessors. This conclusion is consistent with the relatively narrow hip of the enantiornithines and the fact that the diameter of the egg would have been restricted by the size of the pelvic canal, which in these birds was still enclosed by the terminal contact of the two pubic bones.

Various lines of evidence also suggest that enantiornithine hatchlings were highly precocial and capable of finding food on their own. A wealth of fossils of juveniles and even an embryo from the Jehol further indicate that the chicks of these birds hatched covered by a coat of downy-like feathers, and that they became flying fledglings within days after they hatched (some might have been able to fly soon after hatching). Microscopic studies of the characteristics of their bone tissues, often well preserved in their fossils, also indicate that the chicks grew fast during their first year of life. Afterward, they continued growing, but much more slowly over subsequent years. Growth lines visible in paper-thin slices of their bones tell us that, unlike typical living birds, enantiornithines underwent growth cycles interrupted by annual periods in which growth either stopped or was largely arrested until they reached full body size. Also unlike their present-day counterparts, enantiornithine birds were likely able to breed before they reached maximum adult size, as was certainly the case for more primitive birds and their dinosaurian predecessors. This inference, however, would have to be validated by quantitative studies of numerous specimens within the same species; the unprecedented abundance of these birds in the Jehol Biota may soon allow us to test this and other inferences statistically.

The enantiornithines shared a common ancestor with another group of birds that includes all 10,000 living species. Collectively known as ornithuromorphs, the earliest representatives of these birds share a number of features with enantiornithines, characteristics that attest to their close kinship. In the past two decades, a variety of fossils of primitive ornithuromorphs have been unearthed from Early Cretaceous rocks in northern and northeastern China. Other examples of these early cousins of today's birds are also known from Cretaceous sites worldwide, particularly in Argentina, Madagascar, Mongolia, and the United States. These fossils tell us that throughout the Cretaceous,

The grasping design of the foot
of most enantiornithine birds
indicates that many of these
Cretaceous birds lived in trees.

early ornithuromorphs occupied a wide variety of habitats, including deserts, seashores, and floodplains. Some of these birds also became highly specialized swimmers, fully adapted to life in the ocean. The skeletons of the most primitive ornithuromorphs are only slightly different from those of enantiornithines. Overall, however, they display a wide range of appearances and their skeletons show for the first time clear trademarks of their living counterparts.

The majority of these primitive ornithuromorphs were lightly built, flying birds, which tended to be larger than their contemporaneous enantiornithines. Many of them were toothed, although in some cases their teeth were tiny. Several lineages lost their teeth altogether as the result of independent evolutionary events. As in the enantiornithines, the skeletons and plumage of the archaic ornithuromorphs show strong evidence of enhanced aerodynamic capabilities. They had bone proportions in their forelimbs similar to those common among living birds and their shoulder bones were of modern appearance. Some of the early ornithuromorphs also had a powerfully keeled breastbone that provided support for enlarged flight muscles. Nonetheless, enantiornithines and their contemporaneous ornithuromorphs had significant differences in their tail plumage. Unlike enantiornithines, a number of fossils of Cretaceous ornithuromorphs show that the feathered tail of these birds was made up of long shafted feathers that were able to spread out, forming a large airfoil capable of generating aerodynamic forces. The evolution of this fan-shaped tail is considered a milestone in the fine-tuning of avian flight, as this innovation must have provided additional lift and maneuverability.

Characteristics of the bone tissue of the early ornithuromorphs also reveal a degree of modernization in their growth strategies. Details of their bone microstructure show that, like their predecessors, the most primitive ornithuromorphs grew in a cyclical pattern with annual periods in which growth virtually ceased. However, their bone tissue also shows that these birds reached adult size more quickly than their more primitive predecessors, usually within two years. Furthermore, it is in these birds that we witness the origin of the extremely fast growth rates characteristic of present-day birds, which typically reach adult size within the first year of life and prior to becoming sexually active. The bone tissue of Cretaceous ornithuromorphs—*Hesperornis regalis*, *Ichthyornis dispar*, and *Iteravis huchzermeyeri*—which are among the closest cousins of their living relatives show that the distinctive growth and reproductive strategies of modern birds evolved more than 120 million years ago, in species that in other respects were more primitive than those alive today.

The diverse fossils of early ornithuromorphs found in Cretaceous deposits worldwide document the stepwise evolution of many of the features that are typical of their modern kin. The most primitive examples of these birds—the 125-million-year-old *Archaeorhynchus spathula* and *Xinghaiornis lini* from northeastern China—show features in their forelimbs, shoulders, and tails that are in many ways comparable to those of their living relatives. Slightly more advanced forms, such as the hongshanornithids, also known from throughout the Jehol Biota, show a modernized wishbone and breastbone. Later evolutionary stages are well represented by the 80-million-year-old fossils of *Patagopteryx deferrariisi*, excavated from the badlands of Patagonia, in southern Argentina. This flightless, rooster-sized land-dweller exhibits the broad design of the hip (with pubic bones not contacting each other at their tips) that characterizes living birds, an innovation presumably correlated with the capacity to lay larger and less elongate eggs. The evolution of a broader hip (in proportion to the body) might have enabled the evolution

The skeleton and plumage of *Yanornis martini*, an early ornithuromorph from the Jehol Biota, shows the modern proportions of the bones of the forelimb, the advanced design of the shoulder, and the large, fan-shaped tail feathers characteristic of many of these early relatives of present-day birds.

234

of the broader range of hatchling strategies we see among modern birds, but such a transformation might have had a more complex history, perhaps evolving more than once independently.

Further upgrading of the wing skeleton is evidenced in the Mongolian *Apsaravis ukhaana*, which lived 75 million years ago in what is today the Gobi Desert. The wing of this pigeon-sized fossil shows features indicating a capacity to fold and unfold it like that of a living bird. Further advanced ornithuromorphs include the Chinese gansuids and the North American *Ichthyornis dispar*, which lived approximately 120 and 80 million years ago, respectively. The many available fossils of these birds indicate that, while toothed, the skeletons of these close relatives of present-day birds were in many ways similar to those of their living kin. The numerous discoveries of primitive ornithuromorphs of the past few decades have revealed key anatomical and physiological evolutionary events manifested in fossils spanning the last 60 million years of the Cretaceous. These spectacular discoveries have filled an enormous gap in our knowledge about the evolutionary phases that led to the origin of today's birds.

Such a wealth of fossils has also shown that the early relatives of living birds occupied many different habitats and evolved a great variety of lifestyles. Well-preserved fossils of the Jehol ornithuromorph *Yanornis martini* indicate that some of these birds were formidable fish-eaters, presumably foraging far from their nesting areas and carrying food over long distances to feed their chicks. Seeds found inside the crop of relatives of *Hongshanornis longicresta*, a small ornithuromorph also from the Jehol Biota, show that some of these early ornithuromorphs ate grains and stored them in an expansion of the esophagus (the crop). Stomach stones found inside the visceral cavity of a number of these birds indicate that they had also acquired the behavior of ingesting grit that we commonly see among living birds. Their grit-filled, muscular stomachs presumably assisted the digestion of food items in the same way they do among modern birds.

Specializations of the skull and legs of early ornithuromorphs also give us hints about their diverse lifestyles. The long legs and short toes of the 75-million-year-old *Hollanda luceria*, a screamer-sized bird that inhabited the arid regions of what is today the Gobi Desert, suggest that this bird spent most of its time on land. The only known fossil of *Hollanda* is too incomplete to provide definitive evidence about what it ate, but it is possible that it could have been an agile predator that foraged on the many small lizards and mammals that shared its arid habitat. In contrast, the toothless, long beak of Jehol's *Xinghaiornis* suggests that this oystercatcher-sized bird deeply probed the soft mud along the coastline of lakes in pursuit of small invertebrates. Other littoral ornithuromorphs with shorter bills might have pecked just the upper surface of the muddy lakeshore. Among living shorebirds, the length of their bills determines the depth at which they probe or peck muddy or sandy shores in search of food. For example, sandpipers and plovers feed on tiny invertebrates by pecking the surface with their short bills, while godwits, willets, and redshanks thrust their longer bills in search of deeper worms, clams, and tiny crabs.

The preservation of many feeding marks from track sites worldwide have left us with a vivid record of the feeding behaviors of Cretaceous birds—the shape of their footprints suggests that many of the track-makers were primitive ornithuromorphs. Some of these remarkable sites are known in South Korea. Dating back to the first half of the Cretaceous, these Korean track sites show pristine evidence of pecking and probing marks amid thousands of webbed footprints of variously sized birds. These amazing sites even preserve arched bill marks similar to those left today by the alternating sideway motion of spoonbills as they forage on a muddy plain. These and other Cretaceous track sites indicate that many primitive ornithuromorphs—including the long-legged *Archaeornithura meemannae*, the group's oldest known member—foraged along the edge of bodies of water, which just as is common today, hosted a multitude of different birds.

The long toes, and other features of the skeleton of the gansuids (fossils that come from the ancient lake sediments of the Jehol and other 120-million-year-old sites in northern China), indicate that these advanced Early Cretaceous ornithuromorphs were waders and presumably good swimmers. Yet other early ornithuromorphs became far

Clusters of poorly sorted stones concentrated near the pelvic region of many birds from the Jehol Biota demonstrate the presence of a muscular gizzard containing grit, capable of crushing food items.

more specialized for living in the water. Details of the skeletons of these early ornithuro-morphs show that the swimming abilities of gansuids would have been no match against those of the hesperornithiforms, a group of flightless, foot-propelled divers that first appears in the fossil record about 100 million years ago. Initially recognized in the nineteenth century, from discoveries in the famous Niobrara Chalk of western Kansas and other midwestern regions of the United States, these birds are now known from hundreds of fossils, even if most of them are isolated leg bones. Such a rich fossil record comes primarily from a shallow tropical sea that dissected North America during the second half of the Cretaceous, but some of these birds are also known from fossils found in similar ancient environments in Europe and Asia.

Albeit entirely restricted to the aquatic realm, the hesperornithiforms exhibit a rich and diverse evolutionary history straddling 35 million years—their last representatives might have disappeared with the mass extinction that at the end of the Cretaceous wiped out the last of the non-avian dinosaurs. Not much is known about the earliest hesperornithiforms; their fossils indicate they were grebe-sized animals that might have been capable of rudimentary flight. The best-known species is the much younger *Hesperornis regalis*, a 1.2-meter-long (4 feet) formidable diver, larger than an emperor penguin but with puny wings that rendered it entirely flightless. This spectacular animal had a long, narrow head with sharp teeth, an elongate and powerful neck, and a torpedo-shaped body powered by massive legs, all features of a skillful fish-eater. Comprehensive studies of hesperornithiforms show how some of these birds independently evolved large sizes (at least twice) during their long evolutionary history. The fact that as many as four hesperornithiform species—ranging substantially in size—have been found in the Niobrara Chalk of western Kansas also tells us of important differences in how they utilized natural resources, with each species presumably foraging at different depths and feeding on different fish. The diversity of hesperornithiforms shows how early in avian history a variety of birds were able to exploit the marine ecosystems that in the Cretaceous were teeming with life, thus heralding a role that was later played by a cast of other birds, including penguins, grebes, loons, and auks.

Within the family tree of birds, the hesperornithiforms have a place close to their present-day relatives. Another group of near-modern birds, also known from discoveries from the nineteenth century, is represented by the much smaller and flighted *Ichthyornis dispar*. Fossils of this tern-sized, modern-looking bird are often found in the same rocks as those containing the remains of hesperornithiforms, indicating that *Ichthyornis* inhabited the littoral environments of the same shallow seas that were home to its neighbor diver. In some respects, *Ichthyornis* represents a step closer to modern birds than the hesperornithiforms, yet it had sharply toothed jaws designed to catch fish.

Few fossils that can be identified as belonging to groups of living birds have been found in rocks that are 80 to 65 million years old. Because these fossils are for the most part fragmentary, their identification as belonging to lineages of modern birds remains controversial, yet such a record seems to indicate that early cousins of today's shorebirds, ducks, quails, and other familiar birds had already evolved by the end of the Cretaceous. While the fossil record tells us of a small number of modern lineages that seem to have coexisted with some of the archaic groups of Cretaceous birds, genealogies of living birds support a much earlier origin than the age of these modern-looking fossils.

Over the past few decades, DNA researchers have devised clock-like techniques that can estimate the evolutionary origin of the different groups of animals and plants on the basis of the genetic distances among these organisms. Using such techniques, several DNA-based estimates have argued that the earliest relatives of today's living groups of birds must have originated deep in the Cretaceous, more than 100 million years ago. Recent genome-based studies of the evolutionary relationships of living birds have dated the split between the two most basal branches of living birds—ostriches and tinamous and their paleognath relatives from all other birds (a group called neognaths)—about 100 million years ago, and the divergence between landfowl/waterfowl and other neognaths about 88 million years ago. Yet paleontologists have difficulty identifying fossils that are older than 80 million years as belonging to groups of modern birds.

Such a discrepancy between the fossil record and the DNA-based estimates for the origin of modern birds has been mitigated in the past few years with discoveries of abundant fossils of the Chinese gansuids and a refinement of the age of the rocks that

Ichthyornis dispar

Patagopteryx deferrariisi

Hesperornis regalis

The fossil record indicates that by the second half of the Cretaceous, birds had reached an enormous ecological diversity, possibly occupying many of the habitats they occupy today. The roughly coeval (80-million-year-old) *Patagopteryx deferrariisi, Hesperornis regalis,* and *Ichthyornis dispar* illustrate the ecological diversity of Cretaceous ornithuromorphs: *Patagopteryx* was a flightless landbird, *Hesperornis* was an apt sea-diver, and *Ichthyornis* was a flighted bird that lived in nearshore marine environments.

THE EARLY EVOLUTION OF BIRDS

contain them. Because they are both modern looking and closely related to present-day birds, these 120-million-year-old fossils make the molecular predictions a much more plausible scenario. Future discoveries in the Jehol or elsewhere may document the presence of the earliest relatives of modern birds in rocks formed during the first half of the Cretaceous, and hence close the gap that for decades has separated molecular and paleontological interpretations about the timing of the origin of modern birds.

While we are only beginning to determine the ecological role played by the diverse cohort of Mesozoic birds, information from exceptional fossils and from comparisons with their living analogues indicates that many of the lifestyles common to today's birds had already evolved among their Mesozoic forerunners. In fact, the stunning discoveries from the Jehol Biota coupled with fossil findings from around the globe tell us that in contrast to what was once believed, the ecological diversity of the birds from the Mesozoic might have approached the one we see today in their living counterparts. While ancient ecological equivalents of highly specialized groups such as hummingbirds, woodpeckers, and frogmouths have not yet been found, the countless discoveries of Mesozoic-aged fossil birds of the past decades have revealed an impressive and unexpected diversity of ecological adaptations. Clearly, scrutinizing the ancient ecosystems that hosted a multiplicity of birds would have brought no disappointment to the modern-day bird watcher.

The large range of lifestyles evolved by the primitive birds that lived during the Mesozoic included a variety of forms that foraged in lakes and rivers, and others that did so in the ocean. Furthermore, a number of these archaic birds are known to have exploited the wetlands and muddy shores of these bodies of water, their webbed feet and bill marks imprinted in the fossilized mud as they probed the ground in search of food. The diversity of Mesozoic birds also included forms that due to their limited, or even absent, flying capabilities were obligatory land-dwellers, nesting on the ground and feeding on the variety of insects, other invertebrates, and small vertebrates that lived alongside them. Yet many groups of Mesozoic birds were proficient fliers, capable either of making long journeys in search of food or of maneuvering in a close-forested environment. A great number of these birds show features indicating that they lived in trees, eating fruits, seeds, and insects. What we know about the lifestyles of the different birds that lived alongside the extinct Mesozoic dinosaurs speaks of a great deal of diversity, which in some respects heralded the variety of ecological habits present among today's birds.

The ecological milieu of the earliest representatives of modern birds, which lived in the Cretaceous, is not clearly understood. However, the facts that the lineages that are consistently regarded as the closest kin to modern birds have been unearthed from rocks formed in aquatic environments, and that their skeletons show adaptations for living in the water, all suggest that modern birds originated among forerunners that called rivers, lakes, and the ocean their home. It is not known how the early representatives of today's lineages managed to survive the devastating mass extinction that at the end of the Cretaceous decimated many other groups of organisms. Recent genome-based studies have joined other investigations indicating that most lineages of living birds—particularly the supergroup known as Neoaves—diverged around the Cretaceous-Paleocene boundary, the earliest splits of neoavians taking place a few million years before this boundary. These studies indicate that an explosive event of evolutionary divergence took place soon after this boundary, presumably as a result of the availability of ecological niches vacated by the species that succumbed during the environmental turmoil at the very end of the Cretaceous. The fossil record concurs with these genomic studies in showing

that early in the Cenozoic Era, 60 to 50 million years ago, the survivors of these earliest evolutionary divergences diversified into most of the groups of birds that delight us today, carrying along the legacy of the magnificent dinosaurs that ruled the Earth tens of millions of years ago.

AVIAN FLIGHT   Birds are characterized by their ability to move by active flapping flight, a capacity that has enabled them to master mountain ranges and extensive oceans, and to migrate between seasons over other impressive geographic barriers. Flight also allows birds to minimize exposure to predators and provide safe havens to their offspring, and to forage over large territories. This enviable characteristic is without any doubt birds' most remarkable one and most likely the secret of their prodigious success.

The wing of a bird produces four types of basic forces, which interplay with one another during flight. When birds flap their wings, the movement of the air over the wing produces lift as its cambered shape reduces air pressure just above the wing's top surface. Wing flapping also generates forward thrust. These two forces counteract the weight of the bird and the drag that is produced by friction of its body against the air.

Numerous complete fossils of the Early Cretaceous *Gansus zheni* from the Jehol Biota (most likely the same species as the contemporaneous *Iteravis huchzermeyeri*), a bird with remarkably modern-looking skeletal features, have helped ease the controversy between the fossil-based and the DNA-based evidence about the temporal origin of modern birds.

The different flight modes of birds are largely dependent on the overall shape of their wings and bodies, which determines the interplay between force production, maneuverability, speed, efficiency, and other flight characteristics.

Students of bird flight use two important aerodynamic parameters to describe the shape of a wing and to infer the bird's flight performance: aspect ratio and wing loading. Aspect ratio (wingspan/midpoint width) is the relation between the wingspan (the distance from one tip of the wing to the other) and the width of the wing at midpoint. Wings with low aspect ratio (for example, elliptical wings that are short and rounded) are generally adapted for high maneuverability and are common among birds that inhabit forests and other environments with a high density of plants. Many songbirds, which often live among trees and bushes, have elliptical, low aspect ratio wings. Conversely, wings with high aspect ratio are typical of birds that live in environments with minimal vegetation and forage over large distances. Seabirds often have wings that are long and narrow; the tapered wings of fast-flying birds such as falcons, swifts, and swallows also have relatively high aspect ratios. The other key aerodynamic parameter, wing loading (weight/wing area), measures the relation between the weight of a bird and its wing area. Wing loading is usually high among poor fliers, which tend to be heavier and to have relatively small wings. Large fliers with low wing loading—albatrosses, vultures, and others—often fly by gliding or soaring, as these flight styles minimize the energetically costly flapping of the wings.

Collectively, living birds exhibit a wide spectrum of aspect ratios and wing loadings, and their flight styles are best estimated by the combined information from these two key aerodynamic parameters. Most present-day birds have average wing loadings and

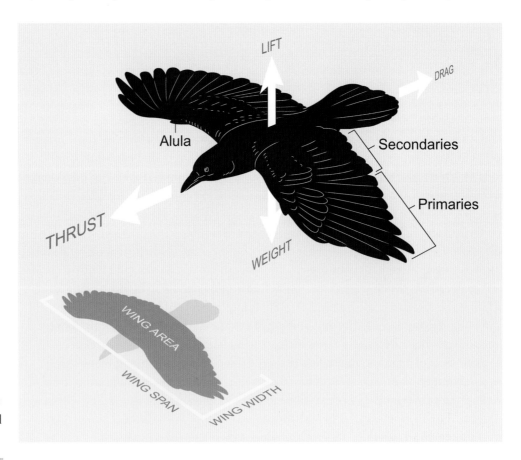

Flying birds are subject to four main forces: thrust and its counterpart (drag), and lift and its counterpart (weight).

aspect ratios; they are good fliers that regularly flap their wings. More specialized fliers are characterized by a distinct relation between these two aerodynamic parameters. Aerial hunters such as swallows, kites, nightjars, and terns have low wing loadings and high aspect ratios that allow them to achieve great speed and maneuverability. A combination of low aspect ratios and low wing loadings is typical of birds that soar high over open areas, such as vultures, condors, eagles, and cranes. Pheasants, turkeys, grouse, and other landfowl have low aspect ratios and high wing loading, which limit their aerodynamic ability to produce short bursts that allow these birds to fly over brief distances. The unique design of the body of hummingbirds, wings with high aspect ratios and a high wing loading, gives these amazing birds their remarkable hovering abilities.

Our ability to estimate the aspect ratio and wing loading of Mesozoic birds has been hampered by the dearth of fossils exhibiting a complete outline of their feathered wings and the challenges associated with inferring the body mass of extinct organisms. Nonetheless, recent statistical analyses of the measurements of the wing bones of modern birds, for which the values for these aerodynamic parameters can be calculated directly, have developed predictive models that can accurately estimate aspect ratios and wing loadings from a suite of skeletal dimensions. When applied to the fossilized remains of the forerunners of modern birds, these advances are providing unprecedented information about the flight styles of pre-modern birds and their non-avian dinosaur relatives.

BECOMING AIRBORNE How the body of birds evolved the architecture that determines their ability to fly and to perform the aerodynamic feats we see every day has been the subject of a long and heated controversy. Historically, this century-old debate has been intertwined with disputes surrounding the ancestry of birds, even if the rise of birds and the beginnings of their flight are two different questions. Today, however, the genealogic knot that for so long tangled the arguments related to the origin of flight in birds has been unraveled: birds, and their flight, evolved from maniraptoran theropod dinosaurs. Additionally, many fossils from around the world have now documented that fundamental flight-related characteristics, such as long, foldable wings and relatively small sizes, evolved prior to the origin of birds, as some lineages of maniraptoran dinosaurs downsized their bodies, increased the length of their feathered forelimbs, and developed a suite of specific structures that allowed movements similar to those we see when birds fly. These fossils have demonstrated that during the evolutionary transition from primitive maniraptorans to birds, lower values of wing loading were achieved as these transitional animals became smaller and their wings became larger. Despite all these great advancements, disagreement about how (and in which ecological context) the flight of birds evolved continues unabated.

For decades since the late 1800s, the debate around the origin of bird flight was polarized between two antagonistic and seemingly incompatible sides. One of these camps argued that birds evolved their flight "from the trees down," a view often referred to as the "arboreal theory." Since its inception, this view has contended that avian flight evolved through a series of incremental stages in tree-climbing animals that leaped between branches, evolved gliding capabilities, and in time developed flapping flight. In the context of a maniraptoran origin of birds, this approach argues that features previously assumed to be predatory specializations of these theropod dinosaurs—long hands with sharply curved claws, among others—are indeed tree-climbing specializations.

The other camp of this historic debate supported a "ground-up" origin of flight, known as the "cursorial theory," which purports that land-dwelling animals involved in the evolutionary transition toward flapping flight never developed arboreal habits. Almost as old as the arboreal theory, the cursorial theory has argued since its inception that avian flight evolved among running animals. Those endorsing this view contend that the ground-dwelling, early predecessors of birds evolved feathered forelimbs capable of producing motions and aerodynamic forces akin to the flight stroke of their present-day descendants. The incipient lift and thrust produced by the "flight stroke" of these animals facilitated leaps among fallen trees, rocks, and other topographic heights, and increased their running speed and ability to climb natural inclines. In time, as these animals reduced their size and increased the aerodynamic performance of their feathered forelimbs, the forces produced by their further refined flight stroke allowed them to take to the air. Those embracing the cursorial theory argue that non-avian maniraptoran theropods did not need to become tree-climbers in order to become airborne, even if it is possible that some early lineages of maniraptorans did take to the trees.

Many details about how the flight of birds evolved in non-avian maniraptoran dinosaurs that were, at least initially, ground-dwellers still remain to be explained, but today some key points of contention have become much clearer. One of these relates to the origin of the flight stroke—the movements of the forelimb that in a bird generate most of its lift and its thrust. As mentioned above, a main argument between the arboreal and cursorial theories was whether such movements evolved before or after the ancestors of birds mastered the ability to become airborne. Basic structures that are widely accepted as necessary to swivel the hand around the wrist, fold the entire forelimb in an avian-like fashion, and sweep it up and down, have been recognized in a variety of early maniraptoran dinosaurs.

Among these structures is a wrist bone in the shape of a half moon with a pulley-like joint surface. A slightly modified version of this bone is also present in modern birds, and studies about how their wrist moves and functions have revealed the important role this bone plays in the folding and unfolding of the hand and the entire wing. Another key aspect of the flight stroke relates to the configuration of the shoulder sockets, which in birds face sideways and enable the wings to sweep up and down as they beat during flight. What we know about the design of the shoulder of some of the maniraptoran predecessors of birds and how this part of the skeleton is situated in exceptionally preserved fossils indicates that these animals had already evolved side-facing shoulder sockets. Such an evolutionary innovation would have permitted the arms of these early, non-avian maniraptorans to swing up and down. Altogether, these and other discoveries indicate that regardless of the context in which the ancestors of birds might have used their arms—for seizing prey, climbing trees, or other purposes—their anatomy reveals that these animals were able to move their forelimbs in a fashion that greatly resembles the flight stroke of living birds.

Another key issue that the fossil record has shed light on in recent years is that feathers with an aerodynamic design—a strong shaft sided by vanes formed by tightly closed barbs—first evolved among ground-dwelling dinosaurs that were clearly not capable of flying and presumably were incapable of climbing trees. The primitive maniraptoran *Caudipteryx*, a member of the group that includes the famous parrot-headed *Oviraptor*, had long, vaned feathers attached to both its arms and tail. The wings of *Caudipteryx* were quite small in comparison to its body—estimates of its wing loading, even if rough approximations, suggest values that are comparable to those of flightless birds. *Caudipteryx* was clearly incapable of flying, but that does not mean it was not able to generate aerodynamic forces with its feathered wings.

Studies of the past decade have highlighted how many land birds—chukar partridges, pheasants, turkeys, and the mound-building megapodes, among others—use aerodynamic forces generated by their wings for purposes other than flying. These birds use the thrust and lift that their wings produce to ascend trees, rocks, or other heights, and to climb vertical walls and overhanging ledges. Such observations have confirmed earlier interpretations based on the fossils of *Caudipteryx* that such primitive maniraptorans could have been capable of generating aerodynamic forces similar to, but still proportionally weaker than, those produced by the wings of flying birds. These early maniraptorans could have used their small, feathered wings either to increase their running speed or to ascend inclines such as rocks or fallen trees.

No specific hypothesis about the origin of avian flight framed within the maniraptoran ancestry of birds can be summarily excluded. Nonetheless, discoveries and interpretations of the past decades have largely mitigated the ecological divide between the traditional arboreal and cursorial theories, and highlighted further how this century-old dispute is in fact a false dichotomy. The available fossil evidence not only reveals that the flight stroke evolved in animals incapable of taking off, but it also tells us that flight itself evolved prior to the origin of birds. During the past 15 years, discoveries of winged paravian theropods from the Jehol and from even older rocks in northeastern China have yielded important information about the flight performance in non-avian dinosaurs.

Earlier we mentioned how several of these early paravians—the troodontid *Anchiornis* and the dromaeosaurid *Changyuraptor*, among others—display the remarkable specialization first documented by the famed *Microraptor*: in addition to having well-formed wings, they also have a set of long, vaned feathers attached to the lower portion of the legs that form a pair of hindwings. Furthermore, the large feathers attached to the forelimbs of these dinosaurs, as well as anatomical details of their shoulders and thorax, leave little doubt about their aerodynamic capabilities. Paramount for assessing the flight competence of these animals is our ability to understand the configuration of their hindwings. Several models of how these animals carried their hindwings have been proposed.

One of these models argues that these feathers projected sideways, attaching perpendicularly along the side of the long axis of the foot bones. In this interpretation, the hindwings would have formed a horizontal airfoil that, together with the forewings, would have resembled the configuration of a biplane. This interpretation, however, makes little sense because, among other things, it is difficult to conceive how these feathers were folded to prevent the obvious damage caused by regular abrasion against branches, foliage, and other objects located at foot level. Additionally, even after careful examination of the position of these feathers in all known fossils, there is nothing indicating that they attached to the foot bones in such a fashion. Conversely, the fossils of these dinosaurs indicate that these feathers attached to the rear margin of the lower leg bones, trailing behind them. Using this more reasonable interpretation of the position of the feathers, another model has argued that *Microraptor* and its kin of hind-winged paravians were capable of sprawling their legs sideways, as if doing the splits, to form a horizontal airfoil with their hindwings. This view, however, is also at odds with what we

At 4 kilograms (almost 9 pounds) and 1.2 meters (4 feet), the dromaeosaurid *Changyuraptor yangi* is the largest known flying non-avian dinosaur. The "hindwings" and long, feathered tail of this animal played a fundamental role in flight control.

know about the fossilized remains of these dinosaurs. The anatomy of the hips and legs of these fossils shows that these animals would have dislocated their legs if they had adopted such a sprawled posture. A more recent model that also favors the reconstruction of the hindwings as attached to the rear side of the legs and trailing after them has argued that these aerodynamic surfaces acted as flight control devices, enhancing the maneuverability of these animals as they banked and moved around in their cluttered forest environment.

The lifestyles of these winged paravians also remain controversial. Ranging from sizes comparable to a large crow, as in the well-known *Microraptor*, to that of the much larger and four times heavier *Changyuraptor*, the strongly arched claws of both hands and feet and the partially opposable first toe of these dromaeosaurids have led some researchers to argue that they lived in the canopy and were capable of climbing trees. The fact that one fossil of *Microraptor* contains the remains of a perching bird in its gut has given some fuel to this view, but other fossils show that this dinosaur also ate fish—so the available evidence from gut contents is inconclusive about the predominant lifestyle of this dinosaur. Other studies have shown little correlation between the anatomy of *Microraptor* and present-day animals that regularly climb trees. A thorough analysis of the skeletal features characteristic of numerous tree-climbing animals— including mammals, lizards, and birds—showed robust statistical support indicating that the anatomy of *Microraptor* and other purported arboreal paravians clusters with that of terrestrial mammals and ground-based birds. Naturally, all this does not mean that some of the dinosaurian cousins of birds could not have ventured into trees, but it does remind us that in most cases we do not have conclusive evidence to say that they did so. Conversely, this and other studies have provided compelling evidence

indicating that the dinosaurian forerunners of birds lived primarily on the ground. Whereas our understanding of more specific aspects of the lives of these animals are still unclear, the design and functional interpretation of the flight-related features of *Microraptor, Changyuraptor,* and its kin of hind-winged paravians provide convincing evidence indicating that some of these forerunners of birds were capable of taking to the air.

Without a doubt, many details about the origin of avian flight remain to be clarified, and the discovery of the bizarre *Yi qi* has added further uncertainty. Yet the available evidence suggests that flight most likely evolved as a byproduct of functions performed by a suite of aerodynamic structures that originated among flightless, land-based dinosaurs. The fossil record tells us that the wing stroke—the key attribute of powered flight—originated among non-avian theropod dinosaurs before they were able to fly, an ability that was achieved only after these animals evolved the smaller sizes and larger wings that allowed them to thrust through the air. The available fossil evidence shows that some of the close cousins of birds, *Microraptor* and other hind-winged paravians, were capable of some sort of aerial locomotion. However, whether this early version of dinosaurian flight—with the intervention of aerodynamic forces produced by hindwings—was the ancestral form of flight inherited by the first birds still remains an open question.

Numerous spectacular fossils of *Microraptor zhaoianus* have been unearthed from the Jehol's 120-million-year-old shales. The characteristics of the plumage and design of the wings of this small dinosaur leave little doubt about its aerial capacity. Nonetheless, the specific flight modes of these animals remain controversial.

FLIGHT REFINEMENT Over the subsequent course of evolution, the refinement of flight involved numerous other transformations, including not only modifications in the skeleton and plumage but also changes in physiology, musculature, and sensory perception. Fortunately, every major lineage of early birds provides evidence that allows us to piece together many of the critical transformations that led to the flight of present-day birds. Indeed, the fossil record tells us of a progressive enhancement of flight performance during much of the Mesozoic evolution of birds, even if different lineages of early birds probably had specific flight modes and degrees of flight competence. Such transformation not only involved the development of key aerodynamic features but also the disassociation of the tail from its role in terrestrial locomotion and an overall escalation in forelimb investment over the that of the hindlimb.

The 150-million-year-old *Archaeopteryx* documents the earliest known phases of avian evolution. This most primitive bird had large wings with flight feathers of comparable appearance and similar in number to the primary and secondary feathers of the wing of many modern birds: it had 12 primaries attached to the hand, 2 more than the 10 primaries typical of most living birds, and about 14 secondaries anchored to the forearm, well within the broad range (6 to 40) of today's birds. *Archaeopteryx* also had a long tail formed by 21 to 23 vertebrae of which all but the first 5 or 6 carried a pair of long, shafted feathers.

Computer tomography of the braincase of one fossil (*Archaeopteryx*'s London specimen) reveals a brain that is comparable in relative size and architecture to those of its non-avian maniraptoran relatives but substantially larger and more complex than those of more primitive dinosaurs. Details of the inner ear also indicate that *Archaeopteryx* was able to move in a three-dimensional space in ways comparable to present-day birds. These studies have shown that *Archaeopteryx* (and its non-avian maniraptoran relatives) had evolved the degree of neurological complexity required for flight.

Relative to the body, the wings of *Archaeopteryx* and other long bony-tailed birds were larger than those of any of their maniraptoran predecessors. The wing loading of the earliest birds was thus lower, a measure that hints at aerodynamic capabilities superior to those of their close dinosaurian relatives. Nonetheless, the overall design of the forelimb of *Archaeopteryx* and other long bony-tailed birds was in many respects much more primitive and quite different from that of their living counterparts. The hand of *Archaeopteryx*, for example, was substantially longer than the bones of the forearm (ulna and radius), and it had three long fingers ending in sharp claws. Such a design was presumably less suitable for generating thrust than the shorter and compact hand of more advanced birds. Furthermore, *Archaeopteryx* lacked an alula, the tuft of small thumb feathers that in modern birds functions as an anti-stalling device when they fly at slow speeds. Recent aerodynamic tests using a life-sized model of *Archaeopteryx* flying in a wind tunnel proved that its long thumb could have functioned as a rudimentary alula, but the fact that this winglet was absent among the most primitive birds highlights once again their aerodynamic limitations.

None of the known specimens of *Archaeopteryx* preserves evidence of a breastbone, which in modern birds provides anchorage for the powerful flight muscles that power the flight stroke. While such muscles could have been anchored to a cartilaginous breastplate, to other bones in the shoulder (particularly, the coracoid), or to a corset of rod-like bones that lined the belly of this animal (gastralia), the apparent absence of a strong breastbone stresses the moderate capacity of its flight musculature. Along the same lines, the more primitive architecture of the shoulder girdle of *Archaeopteryx*, with its wing socket facing sideways, hints that its flight muscles had limitations that precluded the fast, high-amplitude wing beats that are typical of modern birds.

The body plan of *Archaeopteryx* speaks of an animal that might have had an aerodynamic performance similar to that of landfowl, rails, and other heavy fliers; clearly, it was a weaker and less maneuverable flier than most flying birds of today. In fact, new estimations of the wing loading and wing beat frequency of extinct birds based on the statistical relation between these aerodynamic parameters and the proportions of the wing bones in modern birds suggest that *Archaeopteryx* was most likely incapable of any prolonged flapping flight.

The skeleton of *Jeholornis*, while showing signs of modernity in many features, still suggests that this 120-million-year-old bird had many of the same aerodynamic limitations of *Archaeopteryx*. Inferences of key aerodynamic parameters in *Jeholornis* indicate that, like *Archaeopteryx*, this bird was at best able to perform short flights, much like present-day landfowl. Recent studies have also suggested that the long bony tail of these birds played an important role during takeoff and flight control, thus downplaying the contribution of the wings in controlling the aerodynamic trajectories of these animals. The same wind-tunnel experiments that helped explain the function of the thumb have shown that by deflecting its feathered tail, *Archaeopteryx* would have significantly decreased the speed necessary for taking off, thus making it easier (and energetically cheaper) to reach the speed needed for becoming airborne. Other studies suggest that the long-feathered tail of *Archaeopteryx* and *Jeholornis* would have controlled the pitch angle as these birds moved either up or down. Importantly, the role inferred for the long bony tail of these birds implies that the primitive wings of these animals had restrictions for performing the subtle adjustments that are key to the degree of maneuverability we see among living birds. The available information provided by the skeletal architecture, inferred musculature, and brain and ear anatomy of the long bony-tailed birds indicates that while these animals might have been able to fly for short distances, perhaps assisted by an elevated launch, they were unable to achieve the aerodynamic feats characteristic of their living relatives.

A critical step toward the evolution of modern flight is evidenced by the abbreviation of the bony tail into a pygostyle, the bone stump at the end of the vertebral column that in living birds supports the pope's nose (the fatty knob at the rear end of the bird). This notorious transformation most likely heralded a significant enhancement in flight control in the hands of the wings, which in the earliest known, short bony-tailed birds—the Jehol sapeornithids and confuciusornithids—became responsible for the whole set of adjustments necessary for aerial proficiency. Evidence for this is visible in the large wings of these early birds. The ubiquitous *Confuciusornis* had long wings of high aspect ratio that superficially resemble the tapered wings of living terns and swallows; *Sapeornis* had much broader, long wings. The wings of these birds also show an additional skinfold, the propatagium, which connects the wrist with the shoulder. This delta-shaped skinfold, either absent or poorly developed in both *Archaeopteryx* and *Jeholornis*, would have increased the surface area and, hence, the lift-generating capabilities of the wing.

However, it is likely that *Confuciusornis* and *Sapeornis* would have still had severe aerodynamic limitations. Estimations of the wing loading of these birds suggest that *Confuciusornis* would have had wing loadings similar to, or ever greater than, that of *Archaeopteryx*; wing loading estimates for *Sapeornis* are somewhat lower, but just

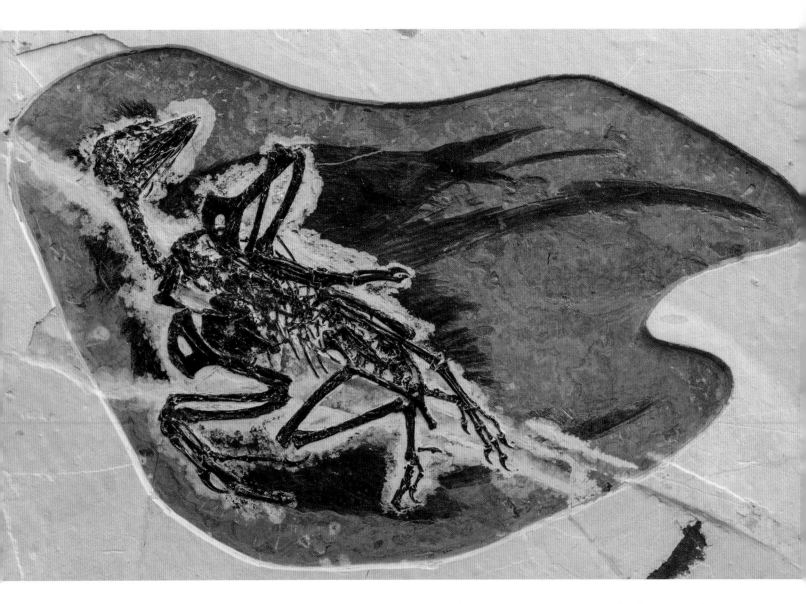

The long and tapered wings of *Confuciusornis sanctus* resemble those of high-speed birds such as swallows, plovers, and terns. Yet, the primitive appearance of the bones of its forelimb and shoulder suggests that this Jehol bird could not have accomplished the aerial feats of its living counterparts. The precise flight styles (and prowess) of the earliest short bony-tailed birds are still poorly understood.

slightly. Yet, the large breastbone of *Confuciusornis* suggests a greater development of its flight muscles when compared to *Archaeopteryx* and the shape of its wings indicate a greater ability to maneuver. While these early short bony-tailed birds are likely to have had flying modes different from those common among their living relatives, the overall evidence from their anatomy, wing profile, and estimated aerodynamic parameters indicates a step toward a fine-tuning in flight capabilities with respect to their long-tailed predecessors.

Many more features associated with the aerodynamic capabilities of living birds are recorded for the first time in the enantiornithines, which in the family tree of birds are a step closer to those alive today. The enantiornithines show a more modern configuration of the wing in which the bones of the forearm form the longest segment of the wing and the fingers are relatively short (two of them are still clawed). Their wings also carry flight feathers—both primaries and secondaries—which are similar in number and arrangement to those of today's flying birds, and their shoulder bones show features that would have imparted more power to the muscles most responsible for the wing stroke. Paramount among these features of the shoulder is a bony passage channeling the tendon of a main muscle, anchored to the breastbone, responsible for the elevation of the wing during the flight stroke.

New discoveries have also revealed that enantiornithines had already developed the same skinfolds that form the fleshy portion of the wing of modern birds, including a complex network of ligaments and tendons that control the movement of the flight feathers. Additionally, the enantiornithine wing displays a well-developed alula, comparable in appearance to that of modern birds. As highlighted before, the alula plays a key role during slow flight as well as during landing and takeoff. This small winglet, attached to the fleshy movable thumb, is located at the midpoint of the leading edge of the wing. It is usually held flush against the wing's edge but when moved forward it creates a gap that separates it from the main wing, an equivalent of a plane's slat. As the air flows through this gap and over the wing, it generates aerodynamic conditions that enable birds to achieve additional lift. In turn, this increased lift allows flying birds to angle their wings into a position that reduces their speed without stalling, a critical maneuver for touching down. The alula thus plays a critical function when birds need additional lift, particularly as they slow down for a safe landing. It also helps birds achieve the necessary lift for taking off from the ground.

Many early enantiornithines are substantially smaller than the most primitive short bony-tailed birds. For example, most Jehol enantiornithines were the size of a typical songbird, while *Confuciusornis* was as large as a crow or a raven and *Sapeornis* might have reached the size of a western gull. The miniaturization experienced by the earliest enantiornithines has a well-understood functional significance: flight performance correlates positively with size reduction. Furthermore, smaller birds—typically with comparatively smaller wing loading—are more maneuverable and flight for them is energetically less expensive. Estimates for the wing loadings of enantiornithines are significantly lower than for those of either *Sapeornis* or *Confuciusornis*, evidence that once again underscores the superior flight competence of the former birds.

One aspect of the aerodynamic quality of these birds not fully understood is the role that their tails played during flight. Enantiornithines had a long, strong pygostyle. However, whether this structure supported a fatty pope's nose—the rectricial bulbs that in modern birds anchor six or so pairs of vaned feathers—is not fully clear. Evidence from fossils of these birds shows that their feathered tail was typically wedge-like and formed by fuzzy, down-like feathers that clearly had no aerodynamic role. In one Jehol bird, the tiny, long-snouted *Shanweiniao cooperorum*, the feathered tail appears to have been formed by a minimum of four large feathers that could have been fanned out into a small, lift-generating surface. A more recent discovery (*Chiappeavis magnapremaxillo*) demonstrates that a fan-shaped feathered tail evolved in at least one enantiornithine. Nonetheless, soft tissues visible around the pygostyle of some exceptionally preserved Jehol enantiornithines show no evidence of the presence of a well-formed pope's nose, thus suggesting that this structure may not have been developed in many of these birds. In spite of this, researchers agree that the advanced skeletal features of enantiornithines, coupled with the presence of an alula, their well-developed wings, and their small size, indicate that even the earliest of these birds, approximately 131 million years ago, had aerodynamic abilities approaching those seen in their extant counterparts.

Many studies have confirmed the close relationship between enantiornithines and ornithuromorphs, the group that includes living birds and their Mesozoic closest relatives. These two groups split from a common ancestor that most likely had already

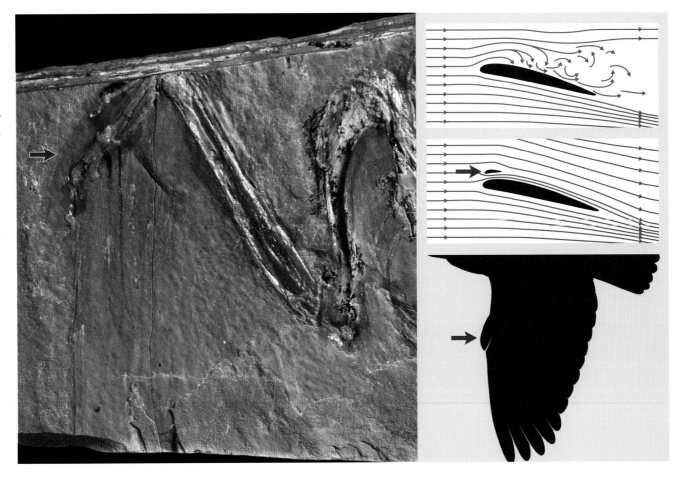

The 115-million-year-old *Eoalulavis hoyasi* from north-central Spain shows a well-developed alula. This small tuft of feathers projecting from the innermost finger plays a key aerodynamic function. When deployed, the alula creates a slot along the wing's leading edge that decreases the turbulence produced by the wing when the bird is flying at low speed. The alula greatly enhances maneuverability and control during takeoffs and landings.

evolved a number of features with advanced aerodynamic functions. Like enantiornithines, the most primitive ornithuromorphs have forelimbs in which the bones have modern proportions: those that form the forearm are the longest and those from the hand form a short, compact unit well suited for generating thrust at the tip of the wing. The architecture of the shoulder girdle of these birds and their large, keeled breastbones indicate that they had powerful flight muscles that would have been able to elevate and depress the wings over a large range of amplitudes. Fossils with outlines of the plumage also show that the wings were large and, at least in some cases, had moderate aspect ratios.

The short pygostyle at the end of the tail of these birds most likely supported a version of the pope's nose, which anchored a fan of shafted tail feathers. Small muscles that adhered to this fatty structure controlled the fanning of these feathers, which in turn provided additional lift and extra control during flight. The tail plumage of these primitive ornithuromorphs has been best studied in the Jehol hongshanornithids, small birds with relatively long legs. Approaching the size of a starling, these birds had wing loadings comparable to those of small, modern birds. The wings of the hongshanornithids were relatively broad but also long and tapered, and their feathered tail was capable of fanning. The proportions of these birds and the characteristics of their plumage indicate that hongshanornithids cruised using a flight mode known as flap-bounding in which birds fly with cycles that alternate between flapping and holding their wings tight against their bodies (bounding). Flap-bounding is a flight mode common to many small living birds; the anatomy of the small hongshanornithids indicates that by at least 125 million years ago, some basal ornithuromorphs had evolved aerodynamic surfaces and flight modes comparable to those of many modern birds of small size.

While they were fully capable of performing acrobatic feats comparable to those we see among today's birds, flight might have been energetically more demanding for these primitive Jehol birds than for their present-day relatives. More advanced birds, closer to the origin of their modern counterparts, show evidence of another important milestone in the transition toward the modern flight stroke. One of these is the 75-million-year-old *Apsaravis ukhaana*, which displays hand and wrist bone details that suggest it was capable of passively unfolding the wing—with minimal muscular intervention—during the upstroke-downstroke cycle. One of these key details is the presence of a prominent attachment for the ligament that supports the propatagium, the triangular skinfold that connects the wrist to the shoulder, expanding the surface area of the wing. This and other features are also visible in the bones of the 85-million-year-old *Ichthyornis dispar*, which like *Apsaravis* is genealogically closer to modern birds than any of the Jehol birds.

Living birds with the same suite of features move their wings by synchronizing the flexion and extension (folding and unfolding, respectively) of the elbow and wrist automatically; the wings fold during the upstroke and unfold during the downstroke with minimal muscular intervention. This remarkable feat is accomplished by a complicated system involving skeletal features, joints, muscles, and ligaments. Some of these features are directly visible in fossils and can be used to infer the development of this automated system in the predecessors of modern birds.

While it is possible that a degree of this automated system, or certain components of it, evolved in earlier forms, the full suite of bony features is already visible in *Apsaravis* and *Ichthyornis*. Such skeletal architecture indicates that at this stage in the fine-tuning of flight, birds were able to operate their wing stroke cycle in a manner similar to that of their living counterparts—folding and unfolding of the wing, with a reduced amount of muscular intervention. Therefore, *Apsaravis* and *Ichthyornis* make clear that at this juncture in the evolution of avian flight, 85 to 75 million years ago, if not before, birds had achieved all the key components of high-performance flapping flight and abilities that were clearly comparable to those of extant birds.

Discoveries of the past few decades have provided us with a paleontological arsenal for deciphering the intricate steps that took place during the long evolutionary fine-tuning of the flight of birds. This fossil bonanza has revealed how a fundamental reduction in body size began among non-avian theropod dinosaurs, many millions of years prior to the rise of birds, and continued during the earliest phases of avian evolution, and how the development of various key structures over millions of years led to the incremental refinement of the aerodynamic toolkit of birds. All this critical evidence has come from fossils greatly separated in time and unearthed across the globe. Yet, no other fossils have given us more information than those from the celebrated Jehol Biota.

THE JEHOL BIOTA At any given time in the deep history of Earth, the fossil record reflects only a small fraction of the myriad of organisms that once lived. Fossils are ubiquitous, but fossilization and preservation are rare phenomena. Animals and plants have a better chance of becoming fossilized if they contain hard parts that can resist a range of destructive forces—currents, scavengers, and bacterial action, among others—which threaten the preservation of any deceased organism. Such preservation potential not only depends on the physical constitution of plants and animals but also on the environment in which their remains become entombed; the best preservation occurs in environments in which organisms are quickly buried by fine sediments and under circumstances that minimize scavenger and bacterial activity. As if these were not enough challenges, fossils also have to survive the mighty geologic forces—from mountain building to erosion—which have ever shaped, and continue to shape, the face of our planet.

While much of the fossil record is largely incomplete due to the rareness of fossilization and the destructive forces that act on the remains of dead animals and plants, here and there a set of exceptional conditions collaborate to mitigate the impact of physical and biological destructive agents and to leave a far more complete record of ancient life. These exceptional events produce fossil assemblages in which organisms are preserved virtually intact, often with soft parts such as feathers, organs, or body outlines still visible in the rock. The sites and rocks containing these rare fossil assemblages are referred to as *Lagerstätten*, a German word that while designating the commercial value of rock quarries has entered the paleontological jargon to distinguish sites characterized by their exceptional preservation. By all means, the fine rocks that preserve the early Cretaceous Jehol Biota of northeastern China are an exquisite example of *Lagerstätten*, with a unique window into a bygone world.

Indeed, the remarkable conditions that have preserved the Jehol Biota have bequeathed us with the clearest perspective on a terrestrial Mesozoic ecosystem and a tremendous wealth of information for understanding the evolution of many groups of animals and plants. In the early part of the Cretaceous, what is today northeastern China was a very different place. Much of this land projected out as an eastward wedge into the ocean. This region was intensively affected by tectonic forces ultimately responsible for the development of a rim of active volcanoes and a network of partially interconnected crater lakes surrounded by temperate forests. It was a time of drastic climatic swings during which the environment became more humid, leading to a significant floral turnover. Life in this verdant world was abundant and diverse, as witnessed by the array of fossils of ferns, ginkgoes, conifers, and flowering plants, and a spectacular menagerie of ancient snails, insects, fishes, frogs, turtles, lizards, flying reptiles, mammals, and birds and other dinosaurs that are all delicately preserved between layers of fine-grained shales.

H. E. Sauvage, a French scientist who studied the fish fauna, undertook the first investigations of the Jehol fossils in the nineteenth century. These initial studies were followed by the work of the American geologist and paleontologist Amadeus W. Grabau, who in association with the Chinese geologists Weng Wenhao and Tan Xichou, studied clam shrimp, fish, insects, and plants from this area in the 1920s. With the Japanese invasion of northeastern China in 1931, the rocks from the Jehol became the focus of study by Japanese researchers, who in the 1930s and 1940s investigated the fossil beds in western Liaoning Province, particularly near the towns of Yixian and Lingyuan, studying the remains of turtles, lizards, and fish from these areas. During the 1950s and 1960s, Chinese researchers continued to study the fossiliferous region. Whereas Grabau had coined the term "Jehol Fauna," it was Zhiwei Gu who in 1962 referred to the fossils exhumed from western Liaoning and adjacent areas as the "Jehol Biota." Chinese paleontologists continued discovering and studying spectacular fossils of insects, plants, and fish during the mid-1970s and 1980s, but a true explosion of discoveries and global interest started in the

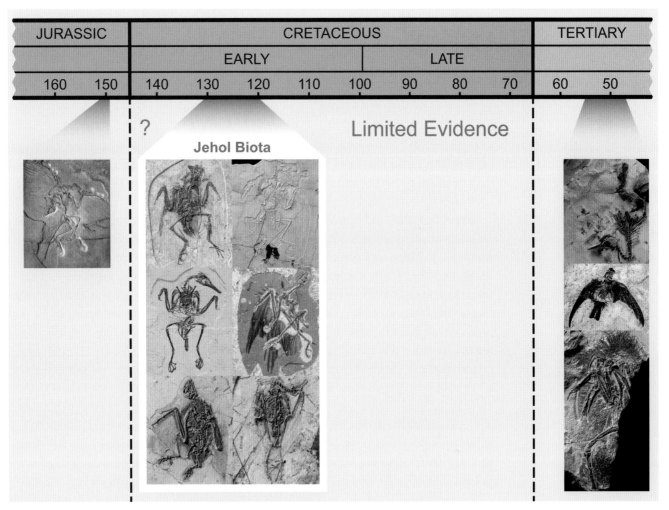

JURASSIC	CRETACEOUS		TERTIARY
	EARLY	LATE	
160 150	140 130 120 110 100	90 80 70	60 50

?

Jehol Biota

Limited Evidence

1990s as the first findings of ancient birds heralded a wave of new fossils, including those of spectacular dinosaurs, mammals, and many other land vertebrates.

As a result of the discovery of many birds and the first feathered dinosaurs, in 1996 the Liaoning Provincial Government approved the creation of a natural protection area for the Beipiao Bird Fossil Area and designated the village of Sihetun, in western Liaoning Province, the core area for special protection. The following year, the Liaoning Geology and Mineral Bureau established a Natural Protection Office for the Beipiao Bird Fossil Area, identifying a large portion of land around Sihetun as a protected zone. In 1998 the State Council of the People's Republic of China designated this as a Natural Protection Area. Today, the worldwide notoriety of these fossils sites continues to grow unabated and, with it, the need to protect them.

The Early Cretaceous lakebeds of western Liaoning Province, and its equivalents along the southeastern margin of Inner Mongolia and the northeastern corner of Hebei Province, are subdivided into three rock units, or geological formations, of multicolored shales and tuffs, the sedimentary layers of which have formed primarily as accumulations at the bottom of lakes. Collectively, these geological formations constitute a thick package of ancient sedimentary layers intermittent with volcanic rocks known as the Jehol Group. The oldest of these geological units is the Huajiying Formation, dated at about 131 million years ago. The layers of fine-grained sediments of this geological

Our knowledge of the early evolutionary history of birds is enlightened by a handful of *Lagerstätten* that have produced many spectacular discoveries of birds. Extensive temporal gaps separating the ages of these sites of exceptional preservation severely limited our understanding of key evolutionary phases in the 150-million-year-old history of birds.

formation are known for containing a variety of crustaceans, insects, fish, and amphibians, among other fossils. Near Fengning, in northeastern Hebei Province, these sedimentary layers have also produced a wealth of fossil birds, the oldest examples of these being animals after the Late Jurassic *Archaeopteryx*. Younger than the Huajiying Formation are the famous Yixian Formation (129 to 122 million years old) and Jiufotang Formation (approximately 122 to 120 million years old); the latter contains the youngest fossils from the Jehol Biota. The shales of the Yixian and Jiufotang Formations have yielded an extraordinary wealth of fossils—the bulk of the Jehol Biota—including amazing examples of most major groups of vertebrates and of many different types of invertebrates. Thus, the combined age of the rocks containing the Jehol Biota has been dated from about 131 to about 120 million years ago.

The Jehol rocks and sites are the best Mesozoic *Lagerstätten*—fossil deposits of exceptional preservation—illustrating what life looked like in a Cretaceous terrestrial ecosystem. At these fossil sites, the remains of ancient life are preserved in gorgeous detail. In many instances, carcasses preserve outlines of the bodies; in other cases, these fossils show a diversity of soft tissues belonging to both internal (muscle bundles, air sacs, ovarian follicles, and others) and external (skin, hair, scales, and feathers) parts of organisms. Several other *Lagerstätten* are known for the Age of Dinosaurs. The Middle Triassic site of Monte San Giorgio along the Swiss-Italian border is famous for the exceptional preservation of a variety of primitive marine reptiles and the ocean life forms that lived along with them. In southern Germany, near the border of France, the Early Jurassic black shales of Holzmaden preserve in exquisite detail the outline of the bodies of many ichthyosaurs and other marine reptiles. The renowned limestone quarries around the village of Solnhofen, in southern Germany, are celebrated as the only place in the world containing the fossils of 150-million-year-old *Archaeopteryx* but, for the most part, the majority of the fossils from these end-Jurassic sites are of fish and marine invertebrates.

The Jehol, however, clearly stands out by preserving fossils of many more animals and plants that lived in lush woodlands than those that lived in the nearby lakes and other bodies of water. Many of the lake inhabitants—aquatic reptiles, amphibians, fish, insect larvae, and other invertebrates—are also preserved in great detail but, unlike any other Mesozoic *Lagerstätten*, the Jehol Biota stands out for the diversity and richness of animals and plants that lived on land. This paleontological bonanza shows a profusion of birds and their dinosaur forerunners that together with a myriad of other animals inhabited massive groves of ginkgoes and conifers. Hundreds of species, and many thousands of exquisite fossils, have been found in the multicolored shales formed at the bottom of the lakes that dotted this wooded region.

Some of the most important fossils of the Jehol Biota are those representing the earliest diversification of flowering plants, one of the most critical evolutionary events in the history of life. With more than 250,000 species, flowering plants represent 90% of all the plants alive today. Their origin and rise to dominance in the Cretaceous transformed the terrestrial ecosystems forever. This significant event, together with an important turnover in the global flora, took place during the 11-million-year interval represented by the Jehol Biota. During this time, the environments dominated by cycads, ferns, and horsetails gradually gave way to forests teeming with conifers, ginkgoes, and flowering plants. Some of the oldest known examples of the latter include the herbaceous *Archaefructus liaoningensis*, an aquatic plant with primitive flowers lacking petals, the fossils of which are abundant in the Jehol rocks. Other flower-bearing herbs are also part of the vast array of plants known from the Jehol Biota. Altogether, these fossils tell us that the Jehol forests were profusely covered by a variety of conifers, including early relatives of yews and other podocarps, pines, cypress, and the monkey-puzzle tree. These forests also contained ginkgoes, seedferns (that carried seeds on frond-like leaves), and a variety of cycads and their kin. Ferns, horsetails, and mosses also formed part of the moist undergrowth and the vegetation closer to the lakeshore.

Different groups of invertebrates, including snails, clams, crayfish, clam shrimps, and spiders, have been found in the Jehol. Nonetheless, the most common invertebrates are insects, the fossils of which document a stunning diversity of both terrestrial and aquatic species, many of them represented by different stages of their lifecycle. In fact, the most abundant fossil in the entire Jehol Biota is the aquatic nymph of a mayfly called *Ephemeropsis trisetalis*. The fine-grained rocks of the Jehol also preserve an amazing number of cockroaches, leafhoppers, beetles, lacewings, cicadas, grasshoppers, midges, and flies, all with delicate structures fossilized in great detail. In fact, hundreds of species of Jehol insects have been named and studied, representing herbivorous, predatory, omnivorous, and parasitic types. Insects, as animals that are highly sensitive to environmental change, provide key evidence for understanding the ancient

Seed plants with leaves superficially resembling those of modern cycads were abundant in the Jehol understory. These and other plants point to stratified forests sustaining a wealth of animal life.

environments of the Jehol. Forest insects are the most common fossils found at these sites. Yet other types of insects provide evidence of the existence of a variety of other habitats, including mountains, lakes, streams, and swamps.

Particularly important is the presence of pollinators. The pollinating activities of these insects can be directly identified by pollen grains preserved in the gut or trapped in the jaws and on the legs of exceptionally preserved fossils, or inferred by their anatomical features, as in the case of the long mouth parts of some insects that are clearly specialized for feeding on deep, tubular flowers. Pollinating insects have played a key role in the early diversification of flowering plants and, as such, in the development of modern environments; the well-studied relationship between the Jehol flowering plant *Archaefructus* and a diversity of pollinating flies living at the same ancient sites documents one of the earliest interactions between flowers and insects.

Virtually every major group of vertebrates has been recorded in the Jehol. Primitive jawless fish are represented by long-snouted freshwater lampreys, the delicate fossils of which add to the poorly known evolutionary history of these early vertebrates. Many fossils of cartilaginous and bony fish have also been exhumed from the Jehol sites, but their diversity is surprisingly limited. This impoverished fossil record includes cartilaginous sturgeons and a variety of primitive ray-finned fish, including archaic teleosts, the group that contains most living fish. Among the latter is the ubiquitous *Lycoptera davidi*,

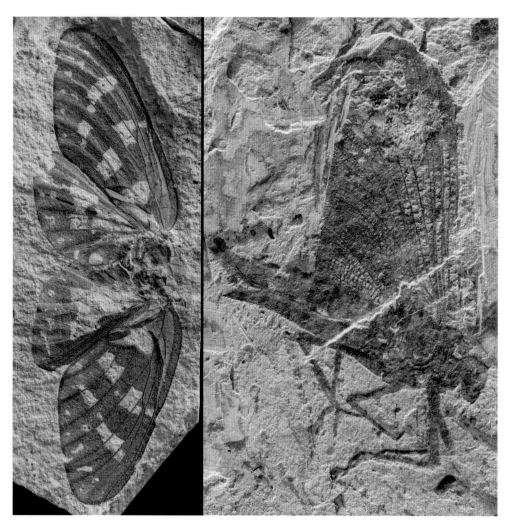

The fine rocks that contain the Jehol Biota have allowed the fossilization of even the most delicate structures with exquisite detail, from the feathers of dinosaurs to the hair of mammals to the wings of insects.

one of the most abundant fossils in the Jehol Biota. A variety of frogs and salamanders also called the Jehol lakes and swamps home. Many of these fossils preserve the outline of the body as well as delicate soft structures such as skin, gills, and eyes. In some sites, the skeletons of these amphibians are preserved in three dimensions, an exception for most sites around the world that yield fossils of these animals.

The remains of numerous frogs and fish from the Jehol provide a glimpse of the many vertebrate animals that inhabited the ancient lakes of this region.

Many virtually intact specimens of the freshwater turtle *Manchurochelys* are preserved in the fine-grained rocks of the Yixian Formation.

Fossils of turtles are abundant but low in species diversity—only a couple of different types of aquatic turtles have been recorded among hosts of other vertebrate species. The curious absence of crocodiles in the Jehol rocks—ubiquitous in concurrent sites around the world—is often used as an argument in support of the cool temperatures envisioned to have prevailed in this region at the time. An extinct group of predatory reptiles, the choristoderans, seem to have taken the ecological niches that elsewhere were occupied by crocodiles. Some of these reptiles had long snouts reminiscent of those of gharials, other Jehol choristoderans had short snouts that made them look more like regular crocodiles, and a third type had extremely long necks and the ability to give birth to live young, which suggest they were fully adapted for life in the water. It is possible that this extinct group of reptiles was able to cope with temperature swings better than their ecological analogues, the crocodiles. While choristoderans had an amphibious to aquatic lifestyle, the Jehol lizards were typically terrestrial. Some of the lizards are thought to have been arboreal, based on the anatomy of their limbs and the strong curvature of their claws. One of them, *Xianglong zhaoi*, possessed ribs interconnected by a membrane capable of being extended sideways as an aerodynamic surface. *Xianglong* was likely able to glide in the same way as the present-day flying dragons.

Another group of reptiles that is remarkably well represented in the Jehol Biota are the pterosaurs, or flying reptiles. The short-tailed pterodactyls dominated the enormous array of pterosaurs known from these sites. Not only did they represent many different branches of the family tree of these archaic flying reptiles, but their fossils also vary greatly in size, from as small as a swallow to others whose wingspans stretched more than 5 meters (close to 17 feet). Most Jehol pterosaurs are thought to have been arboreal animals that fed on fish, a food resource common for pterosaurs worldwide. However, the small, fuzzy-covered *Jeholopterus ninchengensis*, perhaps the most primitive pterosaur from the Jehol, had a broad mouth adapted for feeding on insects, possibly while flying. The extraordinary diversity of forest pterosaurs from the Jehol has transformed our understanding of the ecology of these reptiles. Most pterosaurs from sites around the world have been found in marine sediments and are thus believed to have lived by the sea. Prior to their discovery in the Jehol, the diversity of pterosaurs from inland environments was very reduced and assumed to be evidence of their predominance in coastal settings. Discoveries from the Jehol, however, have forced a revision of this century-old notion.

As mentioned earlier, the Jehol Biota also records an important diversity of dinosaurs. Dozens of species varying greatly in size are represented, and many of them are feathered. As expected from sites that preserve more vividly the actual composition of

the biota, there are fewer carnivorous than herbivorous dinosaurs. The former include early cousins of the mighty *Tyrannosaurus rex* (*Dilong* and *Yutyrannus*), relatives of the renowned *Compsognathus* (*Sinosauropteryx* and *Huaxiagnathus*), and a diversity of small paravians (*Sinornithosaurus*, *Changyuraptor*, *Microraptor*, and many others). Some of these carnivorous dinosaurs, including the compsognathid *Sinosauropteryx* and the dromaeosaurid *Microraptor*, are preserved with the remains of small lizards, mammals, or birds in their guts, which provides direct evidence of their feeding habits. The more common and more diverse herbivorous dinosaurs are represented by the small psittacosaurids (the most common dinosaurs of the Jehol Biota); armored ankylosaurids; distant relatives of duckbills; and the long-necked titanosauriforms, thus far the largest known dinosaurs from the Jehol. Other herbivorous dinosaurs from the Jehol include several species of bizarre theropod groups such as the therizinosaurs, the oviraptorosaurs, and the ornithomimosaurs, which are today considered to have fed largely on plants, even if it is possible that they were omnivores.

The Jehol Biota also includes a variety of primitive mammals, the mouse-sized skeletons of which are often preserved virtually intact, with fur and body outlines. Among these mammals is *Jeholodens jenkinsi*, which lived on the ground and walked with an erect posture, like its modern counterparts. Other Jehol mammals represent

Skeletons of *Sinopterus dongi* together with those of many other pterosaurs—the flying reptiles of the Mesozoic Era—have been unearthed from numerous quarries in Liaoning Province. The diversity of pterosaurs from the Jehol has surpassed that of any other region in the world.

early relatives of the living marsupials and placentals. These fossils provide key information for understanding the rise of these two modern groups, which collectively embody virtually all present-day mammals. The anatomy of the skeletons of most Jehol mammals indicates that they were primarily tree-climbers that lived in forests. Based on the shape of their teeth, they are thought to have eaten mainly insects and worms. Some also included in their diets a healthy dose of plants and seeds, and the formidable *Repenomamus giganticus* ate flesh. With a body comparable to that of a mid-size dog, *Repenomamus* is the largest known mammal of the Mesozoic Era. Some fossils of this ferocious animal preserve the remains of babies of *Psittacosaurus*, a small herbivorous dinosaur, inside the gut. The stunning completeness of the Jehol mammals is in stark contrast to the largely fragmentary fossil record of Mesozoic mammals, which is for the most part represented by isolated teeth and jaw fragments. The discovery of this unexpected diversity of primitive mammals has rewritten many aspects of the early evolution of these animals and has helped us better understand the earliest branches of our own family tree.

The remarkable roster of ancient life preserved in the fine sediments of the Jehol allows us to paint a very complete picture of ecosystems that for millions of years thrived during the first half of the Cretaceous. Evidence from these fossils and information from the rocks that contain them speak of a biota that developed in large lakes connected by streams and in the well-stratified forests that grew around them. This evidence also points to a climate in which the mean annual temperature was around 10 degrees Celsius, where the winters were cold and dry, and the summers were hot and rainy. The rocks and the fossils from these unique sites also tell us that, from time to time, these thriving ecosystems endured episodes of sudden devastation closely linked to the active volcanism that characterized the region. Eruptive emissions of poisonous and pollutant gases,

Pristinely preserved fossils of mouse-sized mammals, such as *Juramaia sinensis*, have been found in the Jehol rocks. Often preserving a halo of fur, most of these fossils have been interpreted as forest-dwellers. Their diversity and distribution across the family tree of mammals, coupled with their exquisite preservation, has transformed our understanding of the early evolution of mammals.

together with large clouds of ash, abruptly changed the environmental conditions on land and altered the chemistry of the lake waters. Such events triggered indiscriminant episodes of mass mortality that decimated life in both the forests and the lakes. Scorching pyroclastic flows occasionally transported the carcasses of land vertebrates to their final burial at the bottom of lakes. Falling ash and other debris from these volcanic eruptions accelerated the burial of corpses, and the low oxygen content of the lake bottom limited the activity of scavengers and decomposing microbes. The geologic agents that led to the final entombment and exceptional preservation of this amazing biota are likely to have been diverse, varying from one specific environment to the other, but the volcanic signature of such carnage is fully evident in the rocks and the fossils of the Jehol Biota.

The Jehol Biota thus presents us with a remarkably clear window into what life, in all its beauty and complexity, was like during the early phases of the Cretaceous, more than 120 million years ago. The ancient drama unfolded by the multihued rocks and their thousands of exquisite fossils is not limited to the intricate ecological interactions between the myriad of organisms that made up these thriving ecosystems. It also includes catastrophic events akin to the renowned disasters endured by many civilizations living in the shadow of deadly volcanoes—from Pompeii to Krakatoa. The awe and carnage left behind has fed our curiosity about this vanished world and helped us answer many questions about the evolutionary history of many groups of organisms. Standing

Countless fossils of the nymph of the mayfly *Ephemeropsis trisetalis*, often forming clusters of hundreds of individuals, have been unearthed from the Jehol quarries. The extraordinary abundance of this fossil is testimony to the mass mortality events that regularly decimated the Jehol Biota.

above all, however, are the insights these magnificent fossils have provided for understanding one of the most fascinating developments in the history of life, that accounting for the evolution of the animals we call birds.

THE JEHOL AVIFAUNA Our understanding of the early evolution of birds is punctuated by the information derived from a handful of fossil *Lagerstätten*—sites of exceptional fossil preservation—which are millions of years apart. *Archaeopteryx*, the earliest fossil bird, is from the famous Solnhofen Limestones of southern Germany, which offers a portrait of the exuberant life in and around a 150-million-year-old tropical lagoon toward the end of the Jurassic when Europe was an archipelago of islands. Millions of years later, the Jehol *Lagerstätten* give us a vivid image of how birds had diversified early in the Cretaceous, some 131 to 120 million years ago in what was at the time a temperate environment of forests developed in between a series of volcanic lakes. Later on, from rocks that are about 55 to 50 million years old, several *Lagerstätten* representing both freshwater and marine environments, in North America and Europe, reveal stunning examples of the diversity reached by the early lineages of modern birds after the mass extinction that closed the Mesozoic Era. These fossils also provide evidence of the rapid burst of diversification and ecological expansion that characterized the evolution of modern birds following the disappearance of their archaic predecessors.

Nonetheless, the prolonged periods of time in between all these rare *Lagerstätten* result in the fact that for most of the evolutionary history of birds, the fossil record is largely incomplete—our knowledge about the birds that lived during those long periods is extremely rudimentary. A consequence of this "*Lagerstätten* effect" is the existence of a 20-million-year gap in the fossil record that separates *Archaeopteryx* from the earliest Jehol birds. Not a single site that can be confidentially dated to this time interval contains the remains of even a single fossil bird. This significant gap in the avian fossil record—as long in duration as the renowned Cambrian Explosion in which most major groups of animals originated early in the Paleozoic Era—has severely hindered our understanding of the opening stages in the evolutionary saga of birds. Macroevolutionary models support the existence of an important surge of anatomical innovation and ecological occupation around the dawn of birds, but critical evidence for understanding such an evolutionary pattern is missing in this 20-million-year gap. The Jehol Biota is our first glimpse into the early diversification of birds that most likely took place at the beginning of the Cretaceous. These exceptional Chinese fossil sites thus open a remarkable window into a crucial juncture in the early history of birds.

The Jehol *Lagerstätten* is also unique in that they record a diversity of birds that straddles almost the entire early portion of their family tree, therefore providing the most comprehensive information about the initial diversification of these animals. Long bony-tailed fossils—close to the ancestry of all birds—are represented by the jeholornithids. Six species of these birds have been named: *Jeholornis prima*, *Jeholornis palmapenis*, *Jeholornis curvipes*, *Dalianraptor cuhe*, *Shenzhouraptor sinensis*, and *Jixiangornis orientalis*. However, more detailed studies are necessary to determine which of these species names are valid. The general consensus among specialists is that most of these birds, if not all of them, belong to a single species, with *Jeholornis prima* as the name most commonly used. An exception is possibly *Jeholornis palmapenis*, which has features of tail plumage that indicate it may be different from *Jeholornis prima*. Nonetheless, very

Often surrounded by a dark halo of fossilized plumage and preserving soft portions of the body, the avian fossils of the Jehol Biota provide paramount information about the early evolutionary history of birds.

subtle anatomical differences separate these two species of *Jeholornis* and additional specimens may prove them to be the result of variability within a single species.

With a size approaching that of a turkey, *Jeholornis* and its kin are among the largest birds known from before the end of the Mesozoic. The skull of these animals was tall and the short jaws bore a few, low-crowned teeth, although the front portion of the upper jaw was toothless. Their forelimbs were primitive, with hands that carried three individualized clawed fingers. The anatomy of their feet suggests that they spent a good portion of their lives on the ground. Their long tail, proportionally longer than that of *Archaeopteryx*, supported a terminal tuft of feathers. These appear to have formed a fan in *Jeholornis prima* and several other named species (*Dalianraptor cuhe* and *Shenzhouraptor sinensis*) and a palm-like arrangement in *Jeholornis palmapenis*, although these differences may well be the result of preservation between the known specimens.

Some fossils suggest that the jeholornithids might have also sported another fan-shaped set of shafted feathers projecting backward from the top of the pelvis, at the tail's base. This configuration has been used to propose that the proximal and distal feather fans (at the base and tip of the tail) might have had different functions—aerodynamic and ornamental, respectively. Although *Jeholornis* had a longer tail than *Archaeopteryx*, important transformations in the bones of the shoulder and the breastbone clearly indicate that it was more closely related to modern birds than *Archaeopteryx*. However, the anatomy of the skeleton of the jeholornithids still points to emergent aerodynamic capabilities when compared to those of more advanced birds, even if the relatively shorter fingers and more abbreviated hand of *Jeholornis* hint at a superior capacity for generating thrust than *Archaeopteryx*.

The Jehol avifauna also contains the earliest known representatives of birds with short bony tails. Among these is the small *Zhongornis haoae*, a species whose avian status, as mentioned earlier, has been recently questioned (although not convincingly). Further studies are needed to document the recent claim that *Zhongornis* is not a bird but a member of the bizarre maniraptoran group that includes the long-handed *Epidexipteryx* and *Yi qi*. As it was originally proposed, *Zhongornis* may well represent the only available evidence of the transition between long bony-tailed birds and their short-tailed counterparts. In this tiny animal, the tail is short but the vertebrae are not fused into a pygostyle, the stump that ends the bony tail of more advanced birds. Therefore, *Zhongornis* may well represent an intermediate stage between the primitive long-tailed birds and those with a bony stump at the end of the tail.

Definitive early members of the short-tailed cohort are well known from the Jehol. The most ubiquitous of them are the toothless confuciusornithids. Known primarily from hundreds of specimens (perhaps thousands) of the crow-sized *Confuciusornis sanctus*, this group of beaked birds also includes a few other named species from the Jehol Biota (some fossils from North Korea may also belong to forms closely related to *Confuciusornis*). The oldest and seemingly most primitive confuciusornithid is *Eoconfuciusornis zhengi*, known from the 131-million-year-old rocks of the Huajiying Formation in Hebei Province, the first sedimentary layers containing fossils from the Jehol Biota. While primarily known by a juvenile specimen, whose skeleton reveals numerous early developmental features, a newly unearthed confuciusornithid from the Huajiying Formation suggests that *Eoconfuciusornis* is indeed different from geologically younger confuciusornithids. Another species clearly different from the ubiquitous *Confuciusornis sanctus* is the smaller *Changchengornis hengdaoziensis*, whose beak was more hooked than that of either *Confuciusornis* or *Eoconfuciusornis*. At about 125 million years old, *Changchengornis* was a contemporary of *Confuciusornis*, both birds being known from the Yixian Formation.

While direct evidence of the diet of the confuciusornithids is still wanting, the massive jaws of these birds—covered by a horny sheath—would have been quite suitable for cracking seeds or equally tough food items. Another distinctive feature of the confuciusornithids is their peculiar hand, which carries long fingers with a unique set of proportions: the outer and innermost fingers end in large and strongly recurved claws, while the middle finger carries a much smaller claw. Some researchers have argued that these birds were capable of climbing trees using their powerfully clawed hands in a crampon-like fashion. However, evidence for a tree-climbing capacity (or even lifestyle) of these birds is lacking. There has never been a compelling functional study that can explain how an animal can use its arms to climb a tree like a squirrel and fly like a bird.

To the contrary, the proportions of the toes and overall structure of the feet suggest that confuciusornithids spent a good amount of time on the ground, presumably foraging. Furthermore, the forelimbs of these birds possess enlarged areas for the attachment of powerful flight muscles, and their large breastbones show evidence of an increased surface for muscle insertion. Coupled with the high aspect ratio of their wings, the available fossil evidence indicates that, when compared to their long-tailed predecessors, the confuciusornithids had developed a step further in aerodynamic competence and degree of flight control. Confuciusornithids did not need to climb trees to take advantage of the canopy; they could simply fly up to them.

The many spectacular specimens of *Confuciusornis sanctus* have given us unprecedented information about a suite of biological characteristics of some of the earliest known birds. In a sense, the sheer numbers of *Confuciusornis sanctus* have turned this fossil into a model organism for early birds, a version of an ancient chicken in which all sorts of studies are undertaken to help our understanding of the lives of primitive birds. An example of these studies involves one of the most salient features of this bird: the highly differential plumage between fossil specimens that are otherwise anatomically indistinguishable. Indeed, some specimens of *Confuciusornis sanctus* carry a pair of long, ribbon-like tail feathers, and others lack them. Detailed examination of these specimens shows that this distinction is not due to differences in preservation but that the presence or absence of these long feathers is a clear characteristic of these animals.

Statistical analyses of hundreds of fossils of *Confuciusornis sanctus*, together with studies of the microstructure of their bone tissue, indicate that the specimens with and without these tail feathers correspond to males and females, respectively. The primary role played by these remarkable feathers is thus interpreted in the context of sex—for display or as ornamentation that attracted females—just like the gaudy feathers found today among many birds. These studies have also revealed that, unlike their present-day counterparts, *Confuciusornis* and its kin began to reproduce before they reached adult size. Like *Archaeopteryx* and its dinosaurian forerunners, early short-tailed birds grew more slowly than their living descendants, which in most instances hatch, fledge, and reach adult body size within the same year. It took several years for primitive birds such as the confuciusornithids to reach adult body sizes. Evidence of this is firmly established by the microstructure of their fossilized bone tissues, which shows growth rings corresponding to annual pauses in bone formation. Such a periodic pattern of growth allows us to estimate the ages at which these animals died and when they began to reproduce.

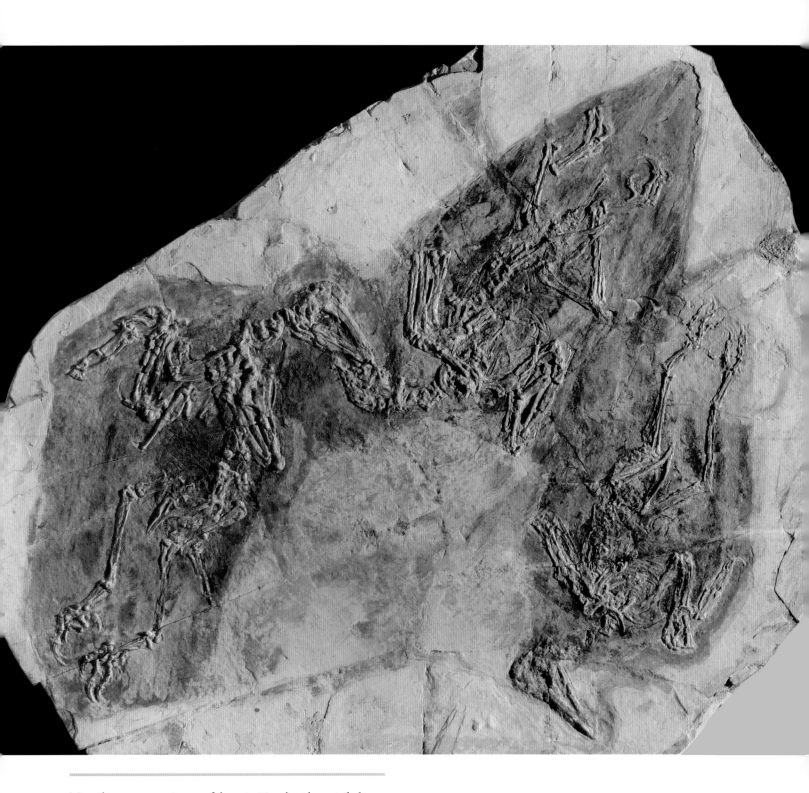

More than 1,000 specimens of the primitive short bony-tailed bird *Confuciusornis sanctus* have been collected from various Jehol quarries. In sites such as Sihetun, fossils of this bird have been unearthed in exceptional concentrations. The frequent association of multiple specimens of this bird is another reminder of the catastrophic demise of many of the inhabitants of the Jehol forests.

Using the presence of ornamental feathers—which, as mentioned earlier, are best interpreted as sexual features—as a proxy for the onset of reproduction, studies show that *Confuciusornis* and its relatives were capable of reproducing when they were still quite young and weighed merely one-quarter that of the largest known specimens.

A similar pattern of growth is known for another primitive group of short-tailed birds, the sapeornithids. Numerous specimens of these birds have been unearthed from both the Yixian and Jiufotang Formations, the sedimentary layers documenting the bulk of the Jehol Biota. Several species of large (*Sapeornis chaoyangensis* and *Didactylornis jii*) and small (*Sapeornis angustis*, *Shenshiornis primita*, and *Omnivoropteryx sinousaorum*) sapeornithids have been named, but detailed studies of the anatomy coupled with the characteristics of the fossilized bone tissue and statistical analyses of their bone proportions suggest they all belong to a single species: *Sapeornis chaoyangensis*.

With a size perhaps comparable to that of a western gull, *Sapeornis* was much larger than any of its contemporaneous short-tailed birds. Its forelimbs were very long, with hands carrying powerful claws on the outer and middle fingers. Evidence from a few specimens preserving portions of plumage indicate that the wingspan of these birds could have reached more than 1.2 meters (4 feet) from tip to tip in the largest specimens. The skull of *Sapeornis* was tall and carried low-crowned, spade-shaped teeth only on the upper jaws; a horny sheath likely protected the lower jaws. The feet of *Sapeornis* had robust toes protected by coarsened pads and ending in sharp claws. The proportions of the toes suggest that these birds were capable of delivering a strong grip, but whether they used their feet to perch on branches or to seize prey is unclear. Nonetheless, evidence from gut contents indicates that *Sapeornis* ate fruits and seeds, which were stored in a crop and crushed inside its muscular gizzard with the help of large grinding stones. The combination of a crop and a gizzard that housed grit for food processing indicates that at this early stage of avian evolution, *Sapeornis* already had developed the main components of the avian digestive system.

There is little doubt that with their large wings, the sapeornithids were capable of taking to the air—how well, however, is not as clear. The broad outline and moderate aspect ratio of their wings suggest that these birds might have been apt gliders, like modern storks and vultures, but detailed studies of their flight are in their infancy. Furthermore, there are aspects of their plumage critical for aerodynamic considerations (in particular, the tail) that are still poorly understood. In addition, the facts that the breastbone appears not to have been ossified and that the shoulder bones have a very unique (and in some ways primitive) design suggest that the wing power, motion, and flight mode of these birds might have been quite different from what we see among living birds of comparable size and wing proportions.

Our understanding of the evolutionary history of enantiornithines, the most species-rich group of Mesozoic birds, greatly hinges on the numerous fossil discoveries from the Jehol Biota. Hundreds of fossils of these birds, belonging to more than 40 species, have been unearthed from the Jehol's sedimentary rocks. The oldest of these fossiliferous layers, the Huayijing Formation of Hebei Province, includes the earliest examples of enantiornithines worldwide. Among these is the starling-sized *Protopteryx fengningensis*, a 131-million-year-old bird known from a handful of skeletons. Another Huayijing enantiornithine is *Eopengornis martini*, a smaller relative of the blunt-toothed *Pengornis houi* from the 120-million-year-old Jiufotang Formation, which, unlike its much younger cousin, had numerous sharp teeth.

Even at this earliest juncture in the evolutionary saga of enantiornithines, *Protopteryx* and *Eopengornis* appear in the fossil record with many of the defining features of this group. Their upper and lower jaws carried small, pointy teeth; their forelimbs had two clawed fingers with bone proportions comparable to those of present-day birds; their wings had an elliptical shape; their shoulder bones had anatomical advances indicating the development of flight muscles capable of powering the wings through the same pulley-like system of modern birds; and their feet were specialized for perching. Fossils of these most primitive enantiornithines signal that early in their history, these birds became specialized for living in trees and maneuvering among dense vegetation.

These birds also display a pair of long ornamental feathers, which attest to the importance of sexual display at even the earliest phases of enantiornithine evolution. Interestingly, these birds greatly differ in the characteristics of their tail feathers. Those of *Protopteryx* exhibit the racket-like design typical of the ornamental feathers of many younger enantiornithines (as well as confuciusornithids), in which a uniformly broad shaft forms most of the feather and well-developed vanes grow only toward the tip. In *Eopengornis*, however, the ornamental feathers have well differentiated vanes throughout their entire length.

The Huayijing sedimentary layers that have yielded the precious fossils of *Protopteryx* and *Eopengornis* also contain the remains of other 131-million-year-old enantiornithines. The discovery of such an array of coexisting species at the time when these birds make their debut in the global fossil record tells us of an older chapter, yet unknown, in the evolutionary history of enantiornithines. Future findings from the beginning of the Cretaceous may be able to unearth evidence detailing the rise of the most successful group of birds from the Mesozoic Era.

The fine sediments containing the Jehol Biota also record other main events that took place early in the evolution of enantiornithines. The stunning variety revealed by their fossils tells us that during these initial phases, these birds increased their spectrum of body sizes and acquired distinct anatomical specializations that together manifest a major expansion of their ecology and lifestyles. The long and small-toothed snout of the tiny *Rapaxavis pani* and its relatives (*Shanweiniao cooperorum* and *Longirostravis hani*) is reminiscent of the narrow bill of plovers and sandpipers. Such adaptation suggests that, like these present-day shorebirds, *Rapaxavis* and its kin used their long snouts to probe the soft substrate of the Jehol lakeshores and their peg-like teeth at the end of their mouths to seize worms and other invertebrates. Their purported relative, *Longipteryx chaoyangensis*, also had a long snout but one that sported a handful of massive and well-recurved teeth. The basic appearance of the snout of *Longipteryx* is more evocative of the serrated bill of the merganser ducks, optimal for catching slippery fish—well-preserved fossils show that the large teeth of *Longipteryx* even carried crenulations along the rear edge that presumably helped the bird seize its prey.

A different group of closely related enantiornithines from the Jehol Biota includes the larger *Bohaiornis guoi* and a half-dozen other species (*Zhouornis hani* and *Parabohaiornis martini*, among them) found in western Liaoning Province. These bohaiornithids are dove- to pigeon-sized birds that lived between 125 and 120 million years ago; they had a full set of strong teeth on both upper and lower jaws, and powerful legs. While it is more difficult to have a better understanding of the particular ecology of these birds, features of their feet have been used to argue that they might have preyed on insects and other small animals. Yet another group of enantiornithines, the

pengornithids, includes the more sizable *Pengornis* (Jehol's largest enantiornithine) and the smaller and much older *Eopengornis*. The small, dome-shaped teeth of *Pengornis*, well-suited for cracking hard food items, and the recurved, sharp teeth of *Eopengornis* provide evidence of a substantial ecological differentiation during the known 11-million-year history of the pengornithids.

Despite the clear adaptations of all these birds for a variety of feeding behaviors, the feet of all Jehol enantiornithines have an overall plan analogous to that of many present-day perching birds; for all intents and purposes, the feet of these ancient birds are like those of birds that spend a considerable amount of time in trees. Ecological characterizations notwithstanding, the fact that the fossil record of Jehol enantiorni-thines shows different birds with distinct ecological specializations provides evidence of how these animals greatly diversified during the early phases of their long evolutionary journey. Additionally, important differences in their wing proportions tell us that early on, these birds evolved an array of different flight styles. While these aerodynamic nuances still need to be untangled, the relatively low aspect ratio of the wings of the Jehol enantiornithines together with the presence of a well-developed alula validate their recognized flight competence, which in general terms might have been compa-rable to that of many small living birds.

Just like in the primitive *Protopteryx* and *Eopengornis*, numerous Jehol enantiorni-thines bear a pair of long ornamental feathers projecting backward from the tail at var-ious lengths—in some instances, they reached more than three times the length of the

All Jehol enantiornithines had teeth on the upper and lower jaws, but their shape, size, and location within the jaws varied considerably. The lower jaws of this bohaiornithid show the stout dentition characteristic of this enantiornithine group.

bird's body. Most typically (but not in *Eopengornis*), these feathers are characterized by having a very broad shaft, which gives them a strap-like appearance, and a rounded end formed by two vanes of differently sized barbs. Their specific shape differs from that of any modern feather; these ancient feathers are essentially extinct. Yet, back in the Cretaceous, they ornamented the tails of many enantiornithine species. These tail feathers had different sides and overall appearance depending on the species, attributes that, together with their gaudiness, tells us that they had the purpose of attracting mates. The evolution of ornamental traits such as the elaborate plumage of the peacock, or the quetzal of tropical Latin America, is often explained as the result of processes acting on features that have a sexual function, such as courtship. It is assumed that these ornaments, most typical of males, signal females that the most flamboyant males have the best physical qualities. The existence of many of these outstanding plumages among even the earliest enantiornithines tells us that these primitive birds had mating rituals akin to those we see today among many of their counterparts.

The Jehol enantiornithines have also illuminated other aspects of the biology of early birds. Many of these fossils are juvenile birds, including an embryo still contained within its egg. Evidence from these fossils signals that enantiornithine chicks were relatively independent and capable of foraging from the time they hatched. Studies of their bone tissue—often preserved just as exquisitely as their skeletons—point to rapid growth rates when they were young, with a marked decrease in the pace of bone formation their first year. Nonetheless, the characteristics of their bone tissue indicate that these animals continued to grow over several years and had annual pauses in which their growth was greatly, if not entirely, curtailed.

As in more primitive birds (*Confuciusornis*, among others), and unlike in living birds, enantiornithines appear to have had their first breeding season before becoming full-grown. This primitive type of reproductive strategy, more characteristic of what we see among present-day reptiles than living birds, is in contrast with what we know about how these birds laid their eggs. A handful of rare examples of Jehol fossils suggest that, as in present-day birds, enantiornithines may have had a single functional oviduct, which laid one egg at a time (presumably every day or every few days). Evidence of this comes from tantalizing discoveries of ovarian follicles—cellular aggregations that contain the immature egg—inside the body cavity of some fossils. These spectacular discoveries have not been exempt from controversy; some researchers have expressed skepticism about these structures being fossilized follicles, primarily claiming that the preservation of such delicate tissues would be highly exceptional. Nonetheless, while indeed an apparent exceptional case of preservation, the bulk of the evidence does lean toward the interpretation of these structures as ovarian follicles. If properly interpreted, these findings would indicate that physiologically, even the earliest known enantiornithines were in some ways similar to their living counterparts.

In the past decade, a great number of new fossils representing the early phases of evolution of the ornithuromorphs—the group that includes all living birds and their closest relatives—have been unearthed from the Jehol sediments. The available information about these birds point to marked ecological differences with respect to coexisting enantiornithines. Body mass estimates show that early ornithuromorphs were generally larger than their enantiornithine counterparts. Basic differences in the anatomy of their feet also point to the types of environments these two groups of birds inhabited. Whereas the structure of the feet of most Jehol enantiornithines supports an arboreal existence, a life

Most Jehol enantiornithines had body dimensions similar to those of small songbirds; others, like this bohaiornithid, were about the size of a pigeon.

intimately connected to the Jehol forest, the anatomy of the feet of many of their contemporaneous ornithuromorphs indicate that these birds had either a more terrestrial lifestyle or specializations for living at the lakeshore and in the water.

As in the case of other early avian groups, the Jehol Biota has yielded the remains of the earliest known ornithuromorph: the 131-million-year-old *Archaeornithura meemannae*. The anatomy of the skeleton of this plover-sized bird identifies it as an early member of the hongshanornithids, a group of long-legged, small birds well known from younger Jehol stratigraphic layers. While the hongshanornithids are the oldest recognized ornithuromorphs, they are not necessarily the most primitive within this bird group. Ornithuromorphs that are further removed from the ancestry of modern birds include *Archaeorhynchus spathula*, a quail-sized, beaked animal with elongated wings and stout feet, and *Xinghaiornis lini*, another toothless bird, albeit larger and with a long, slender bill. The skeletons of these most primitive ornithuromorphs show features that are also common to enantiornithines, thus indicating a proximity to the evolutionary split between ornithuromorphs and enantiornithines. Although the overall design of the wings of early ornithuromorphs is similar in proportions to that of the enantiornithines, the wing bones of archaic ornithuromorphs look much more like those of living birds in a number of details. Their shoulder bones also have an essentially modern configuration, and the breastbone carries a larger keel for the attachment of powerful flight muscles.

Other differences with respect to more primitive birds—enantiornithines included—are evidenced in the tail of many Jehol ornithuromorphs, particularly in the shape of the pygostyle (the rump bone at the end of the tail), which became much shortened, approaching that of present-day birds in overall appearance. The modifications in the tail of these early ornithuromorphs are probably related to the evolution of a modern type of rectricial bulbs (the structures forming the fatty end of the tail known as the

Miniscule teeth in the skull of the Jehol gansuid *Gansus zheni* illustrate a trend toward tooth reduction characteristic of many early lineages of ornithuromorph birds.

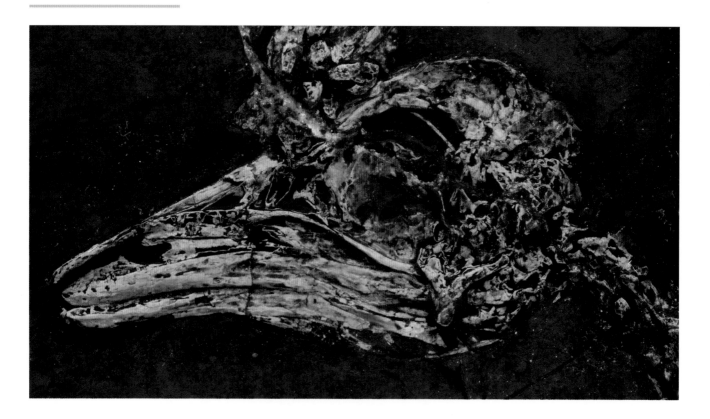

pope's nose), which developed into a broad anchor for a set of long, vaned feathers (usually six pairs) capable of fanning out. The fan-shaped feather tail of living birds is known to maximize lift when fully spread, and some Jehol fossils tell us that the tails of early ornithuromorphs were formed by a similar number of vaned feathers capable of fanning to degrees comparable to those we see among today's birds. Once again, evidence from the spectacular Jehol fossils indicates that as early as 131 million years ago, birds had evolved critical aerodynamic features, such as a fan-shaped feather tail, which contributed to their flight competence.

The contrasting specializations of *Archaeorhynchus* and *Xinghaiornis* underscore a significant degree of ecological differentiation at the onset of ornithuromorph evolution. The stout feet and relatively short beak of *Archaeorhynchus* suggest that this was primarily a ground-dweller that presumably foraged on the forest floor while concealed by the undergrowth. In contrast, the relatively long legs of *Xinghaiornis* allude to its wading habits, and its long and slender bill tell us that it most likely fed on the muddy lake edges, probing and pecking in search of worms and other invertebrates. The large spectrum of sizes of the early ornithuromorphs, for example, between two lineages of waders such as the plover-sized hongshanornithids and the much larger whimbrel-sized *Xinghaiornis*, further emphasizes the ecological differences that clearly existed among many of these birds. These differences are also accentuated by the marked anatomical variation of their skulls and teeth, which span from toothless forms (*Archaeorhynchus* and *Xinghaiornis*, among others) or animals with minute dentition (hongshanornithids and gansuids) to others that retained powerful teeth throughout their mouths, particularly those that fed on fish. The best known of these fish-eating, archaic ornithuromorphs is the tern-sized *Yanornis martini*, whose skull had a relatively long snout with many sharply pointed teeth on both the upper and lower jaws.

The elongate jaws of the Jehol ornithuromorph *Yanornis martini* bore a large quantity of closely spaced, sharp teeth. Numerous fossils of this bird are preserved with the remains of small fish (sometimes whole) inside the crop and digestive tract. The sharp teeth of this bird would have helped it secure the slippery fish it ate.

The diversity of Jehol ornithuromorphs spreads over a large portion of the family tree of these birds, thus revealing several stages in the acquisition of the body plan of modern birds. The primitive *Archaeorhynchus* and *Xinghaiornis* are clearly close to the origin of the ornithuromorphs, but other Jehol fossils clearly belong to more advanced lineages, evolutionarily closer to their modern counterparts. In fact, the anatomy of the hongshanornithids, *Yanornis*, and other ornithuromorphs from the Jehol provides evidence of an escalation toward the overall skeletal plan of present-day birds. The greatest degree of modernity is represented by the gansuids, grebe-sized birds with long legs that probably inhabited the shores and calm waters of ancient lakes. Gansuids were once distributed across northern China—many fossils of two species, *Gansus yumenensis* and *Gansus zheni* (possibly the same as *Iteravis huchzermeyeri*), have been unearthed in quarries dating back to 120 million years ago in Gansu and Liaoning Provinces, respectively. These birds were remarkably modern in many details of their skeletons but unlike their living counterparts, had tiny teeth they possibly used for securing lakeshore invertebrates.

In recent years, greater understanding of the anatomy and age of these birds has helped mitigate a decades-long debate between paleontologists and molecular biologists about when precisely the group that includes all modern birds originated. We have pointed out how the earliest record of birds—the celebrated *Archaeopteryx*—is approximately 150 million years old, but the record of birds identified as belonging to modern lineages (ducks, geese, grouse, pheasants, and so on) is limited to fossils occurring in rocks that are younger than 70 to 80 million years old. This evidence notwithstanding, studies analyzing differences in the sequence of DNA molecules among present-day birds suggest a much older origin for the group that includes all living birds. Armed with the powerful evidence of the genetic makeup of modern birds, molecular biologists have consistently argued that this group evolved deep in the Cretaceous, more than 100 million years ago. This multimillion-year discrepancy between what the fossils document and what the genes signal has sparked heated debates over the past few decades.

The spectacular fossils of gansuids help bridge the divide between these two lines of evidence. On the one hand, these birds are anatomically very modern and clearly, closely related to modern birds; on the other hand, they existed 40 to 50 million years prior to the earliest records of present-day lineages. Thus, by embodying both anatomical modernity and stratigraphic antiquity, the gansuids have mitigated the controversy between the molecular and the paleontological evidence like no other discovery. These remarkable fossils tell us that the evolutionary history of modern birds most likely began during the first half of the Cretaceous, many millions of years before the oldest known records of these birds.

The Jehol Biota has boosted our understanding of the beginnings of birds like no other evidence. The exceptional preservation of the fossil birds from these sites and their impressive diversity have provided unprecedented evidence of almost every chapter of the early evolutionary history of birds. As a matter of fact, close to one-third of the diversity of Mesozoic birds worldwide has been unearthed from these fossil-laden rock layers. The stunning fossils from these ancient sites paint a picture in which very different birds lived side by side with, and flew over, their dinosaurian forerunners. Each of these fossils recounts a page of the long chronicle of avian evolution, from primitive long-tailed birds with rudimentary flight capabilities to others that looked much more modern, with a keen degree of flight competence and approaching evolutionary stages close to present-day birds.

Numerous small enantiornithines, often preserving the contour of their wings and overall plumage, have been found in the Jehol rocks. While many detailed studies of these birds are yet to be conducted, the wealth of information provided by these fossils is unparalleled in the history of paleornithology.

Preserving virtually intact feathers, accurate contours of wings and tails, a myriad of other soft structures, and all stages in the lifecycle of these animals, these exceptional fossils provide an incomparable source for interpreting the lives of early birds and their epic evolutionary journey. They speak of an initial surge of diversification leading to many different lifestyles, mirroring what we see among their living counterparts. They fill the enormous evolutionary gap between birds of modern pedigree and the formidable dinosaurs that gave rise to them. They narrate a multimillion-year story of evolution, in which most of the key adaptations of modern-day birds became fine-tuned and took on their current shape. In the end, the Jehol birds provide us with unparalleled evidence for understanding how living birds became the animals they are. Three decades after a split rock revealed the first of thousands of skeletons at a quarry in northeastern China, these remarkable birds of stone evoke an ancient cemetery in which numerous fossils, like intricate tombstones, speak of how some of the earliest birds lived and died, and how their stories connect the lives of the long-extinct dinosaurs with the avian menagerie of today.

BIBLIOGRAPHY

Chang, Mee-mann, Pei-ji Chen, Yuan-quing Wang, Yuan Wang, and De-sui Miao. *The Jehol Fossils: The Emergence of Feathered Dinosaurs, Beaked Birds and Flowering Plants.* London: Academic Press, 2008.

Chiappe, Luis M. *Glorified Dinosaurs: The Origin and Early Evolution of Birds.* Hoboken, NJ: John Wiley, 2007.

Chiappe, Luis M., and Lawrence M. Witmer. *Mesozoic Birds above the Heads of Dinosaurs.* Berkeley: University of California Press, 2002.

Dingus, Lowell, and Timothy Rowe. *The Mistaken Extinction: Dinosaur Evolution and the Origin of Birds.* New York: W. H. Freeman, 1998.

Gill, Frank B. *Ornithology.* 3rd ed. New York: W. H. Freeman, 2007.

Kaiser, Gary W. *The Inner Bird: Anatomy and Evolution.* Vancouver: UBC Press, 2007.

Long, John A., and Peter Schouten. *Feathered Dinosaurs: The Origin of Birds.* Oxford: Oxford University Press, 2008.

Padian, Kevin, and Luis M. Chiappe. "The Origin of Birds and Their Flight." *Scientific American*, February 1998, 38–47.

Pickrell, John. *Flying Dinosaurs: How Fearsome Reptiles Became Birds.* New York: Columbia University Press, 2014.

Proctor, Noble S., and Patrick J. Lynch. *Manual of Ornithology: Avian Structure and Function.* New Haven: Yale University Press, 1998.

Shipman, Pat. *Taking Wing: Archaeopteryx and the Evolution of Bird Flight.* New York: Simon & Schuster, 1998.

THE WORLD OF THE MESOZOIC

Levin, Harold L. *The Earth through Time.* Philadelphia: Saunders, 1978.

Okada, Hakuyu, and Niall J. Matteer, eds. *Cretaceous Environments of Asia.* Amsterdam: Elsevier, 2000.

Steuber, Thomas, Markus Rauch, Jean-Pierre Masse, Joris Graaf, and Matthias Malkoč. "Low-latitude Seasonality of Cretaceous Temperatures in Warm and Cold Episodes." *Nature*, no. 437 (2005): 1341–1344.

MESOZOIC LIFE

Benton, Michael J. *Vertebrate Palaeontology.* 3rd ed. Malden, MA: Blackwell Science, 2005.

Brett-Surman, M. K., Thomas R. Holtz Jr., and James O. Farlow, eds. *The Complete Dinosaur.* 2nd ed. Bloomington: Indiana University Press, 2012.

Brusatte, Stephen L., Sterling J. Nesbitt, Randall B. Irmis, Richard J. Butler, Michael J. Benton, and Mark A. Norell. "The Origin and Early Radiation of Dinosaurs." *Earth-Science Reviews*, no. 101 (2010): 68–100.

Sereno, Paul C. "The Origin and Evolution of Dinosaurs." *Annual Review of Earth and Planetary Sciences*, no. 25 (1997): 435–589.

Weishampel, David B., Peter Dodson, and Halszka Osmólska, eds. *The Dinosauria.* 2nd ed. Berkeley: University of California Press, 2007.

Balanoff, Amy, Gabe S. Bever, Timothy B. Rowe, and Mark A. Norell. "Evolutionary Origins of the Avian Brain." *Nature* 501, no. 7465 (2013): 93–96.

Bhullar, Bhart-Anjan S., Jesús Marugán-Lobón, Fernando Racimo, Gabe S. Bever, Timothy B. Rowe, Mark A. Norell, and Arhat Abzhanov. "Birds Have Paedomorphic Dinosaur Skulls." *Nature*, no. 7406 (2012): 223–226.

Chiappe, Luis M. "Downsized Dinosaurs: The Evolutionary Transition to Modern Birds." *Evolution: Education and Outreach*, no. 2 (2009): 248–256.

Clark, James M., Mark Norell, and Luis M. Chiappe. "An Oviraptorid Skeleton from the Late Cretaceous of Ukhaa Tolgod, Mongolia, Preserved in an Avianlike Brooding Position over an Oviraptorid Nest." *American Museum Novitates*, no. 3265 (1999): 1–36.

Erickson, Gregory M., Oliver W. M. Rauhut, Zhonghe Zhou, Alan H. Turner, Brian D. Inouye, Dongyu Hu, Mark A. Norell, and Robert Desalle. "Was Dinosaurian Physiology Inherited by Birds? Reconciling Slow Growth in *Archaeopteryx*." *PLoS ONE* (2009): DOI:10.1371/journal.pone.0007390.

Lee, Michael S. Y., Andrea Cau, Darren Naish, and Gareth J. Dyke. "Sustained Miniaturization and Anatomical Innovation in the Dinosaurian Ancestors of Birds." *Science* 345, no. 6196 (2014): 562–566.

Mackovicky, Peter J., and Lindsay E. Zanno. "Theropod Diversity and the Refinement of Avian Characteristics." In *Living Dinosaurs: The Evolutionary History of Modern Birds*, edited by Gareth Dyke and Gary Kaiser, 9–29. Hoboken, NJ: John Wiley, 2011.

Organ, Chris L., Andrew M. Shedlock, Andrew Meade, Mark Pagel, and Scott V. Edwards. "Origin of Avian Genome Size and Structure in Non-avian Dinosaurs." *Nature*, no. 446 (2007): 180–184.

Ostrom, John H. "*Archaeopteryx* and the Origin of Birds." *Biological Journal of the Linnean Society* 8, no. 2 (1976): 91–182.

Padian, Kevin, and Luis M. Chiappe. "The Origin and Early Evolution of Birds." *Biological Reviews of the Cambridge Philosophical Society* 73, no. 1 (1998): 1–42.

Sumida, Stuart S., and Christopher A. Brochu. "Phylogenetic Context for the Origin of Feathers." *Integrative and Comparative Biology* 40, no. 4 (2000): 486–503.

Xu, Xing, Hailu You, Kai Du, and Fenglu Han. "An *Archaeopteryx*-like Theropod from China and the Origin of Avialae." *Nature* 475, no. 7357 (2011): 465–470.

Xu, Xing, Zhonghe Zhou, Robert Dudley, Susan Mackem, Cheng-Ming Chuong, Gregory M. Erickson, and David J. Varricchio. "An Integrative Approach to Understanding Bird Origins." *Science* 346, no. 6215 (2014): 1341.

FEATHERED DINOSAURS

Hone, David W. E., Helmut Tischlinger, Xing Xu, Fucheng Zhang, and Andrew Allen Farke. "The Extent of the Preserved Feathers on the Four-Winged Dinosaur *Microraptor gui* under Ultraviolet Light." *PLoS ONE* (2010): DOI:10.1371/journal.pone.0009223.

Hu, Dongyu, Lianhai Hou, Lijun Zhang, and Xing Xu. "A Pre-*Archaeopteryx* Troodontid Theropod from China with Long Feathers on the Metatarsus." *Nature*, no. 461 (2009): 640–643.

Ji, Qiang, Philip J. Currie, Mark A. Norell, and Ji Shu-An. "Two Feathered Dinosaurs from Northeastern China." *Nature*, no. 393 (1998): 753–761.

Norell, Mark A., and Xing Xu. "Feathered Dinosaurs." *Annual Review of Earth and Planetary Sciences* (2005): 277–299.

Xu, Xing. "Feathered Dinosaurs from China and the Evolution of Major Avian Characters." *Integrative Zoology* 1, no. 1 (2006): 4–11.

Xu, Xing, Kebai Wang, Ke Zhang, Qingyu Ma, Lida Xing, Corwin Sullivan, Dongyu Hu, Shuqing Cheng, and Shuo Wang. "A Gigantic Feathered Dinosaur from the Lower Cretaceous of China." *Nature* 484, no. 7392 (2012): 92–95.

Zhang, Fucheng, Zhonghe Zhou, Xing Xu, Xiaolin Wang, and Corwin Sullivan. "A Bizarre Jurassic Maniraptoran from China with Elongate Ribbon-like Feathers." *Nature*, no. 455 (2008): 1105–1108.

FLYING DINOSAURS

Han, Gang, Luis M. Chiappe, Shu-an Ji, Michael Habib, Alan H. Turner, Anusuya Chinsamy, Xueling Liu, and Lizhou Han. "A New Raptorial Dinosaur with Exceptionally Long Feathering Provides Insights into Dromaeosaurid Flight Performance." *Nature Communications* 5 (2014): DOI:10.1038/ncomms5382.

Xu, Xing, Xiaoting Zheng, Corwin Sullivan, Xiaoli Wang, Lida Xing, Yan Wang, Xiaomei Zhang, Jingmai K. O'Connor, Fucheng Zhang, and Yanhong Pan. "A Bizarre Jurassic Maniraptoran Theropod with Preserved Evidence of Membranous Wings." *Nature* 521, no. 7550 (2015): 70–73.

Xu, Xing, Zhonghe Zhou, Xiaolin Wang, Xuewen Kuang, Fucheng Zhang, and Xiangke Du. "Four-winged Dinosaurs from China." *Nature*, no. 421 (2003): 335–340.

THE ORIGIN OF FEATHERS

Brush, Alan H. "Evolving a Protofeather and Feather Diversity." *Integrative and Comparative Biology* 40, no. 4 (2000): 631–639.

Li, Quanguo, Julia A. Clarke, Ke-Qin Gao, Chang-Fu Zhou, Qingjin Meng, Daliang Li, Liliana D'Alba, and Matthew D. Shawkey. "Melanosome Evolution Indicates a Key Physiological Shift within Feathered Dinosaurs." *Nature* 507, no. 7492 (2014): 350–353.

Li, Quanguo, Ke-Qin Gao, Jakob Vinther, Matthew D. Shawkey, Julia A. Clarke, Liliana D'alba, Qingjin Meng, Derek E. G. Briggs, and Richard O. Prum. "Plumage Color Patterns of an Extinct Dinosaur." *Science* 327, no. 5971 (2010): 1369–1372.

Manning, Phillip L., Nicholas P. Edwards, Roy A. Wogelius, Uwe Bergmann, Holly E. Barden, Peter L. Larson, Daniela Schwarz-Wings, Victoria M. Egerton, Dimosthenis Sokaras, Roberto A. Mori, and William I. Sellers. "Synchrotron-based Chemical Imaging Reveals Plumage Patterns in a 150 Million Year Old Early Bird." *Journal of Analytical Atomic Spectrometry*, no. 28 (2013): 1024–1030.

Prum, Richard O. "Evolution of the Morphological Innovations of Feathers." *Journal of Experimental Zoology Part B: Molecular and Developmental Evolution* 6, no. 304 (2005): 570–579.

Stettenheim, Peter R. "The Integumentary Morphology of Modern Birds—An Overview." *Integrative and Comparative Biology* 4, no. 40 (2000): 461–477.

Wogelius, R. A., P. L. Manning, H. E. Barden, N. P. Edwards, S. M. Webb, W. I. Sellers, K. G. Taylor, P. L. Larson, P. Dodson, H. You, L. Da-Qing, and U. Bergmann. "Trace Metals as Biomarkers for Eumelanin Pigment in the Fossil Record." *Science* 333, no. 6049 (2011): 1622–1626.

THE MESOZOIC AVIARY

Bell, Alyssa K., Luis M. Chiappe, Gregory M. Erickson, Shigeru Suzuki, Mahito Watanabe, Rinchen Barsbold, and Khishigjav Tsogtbaatar. "Description and Ecologic Analysis of *Hollanda luceria*, a Late Cretaceous Bird from the Gobi Desert (Mongolia)." *Cretaceous Research* 31, no. 1 (2010): 16–26.

Benson, Roger B. J., and Jonah N. Choiniere. "Rates of Dinosaur Limb Evolution Provide Evidence for Exceptional Radiation in Mesozoic Birds." *Proceedings of the Royal Society B: Biological Sciences* 280, no. 1768 (2013): DOI:10.1098/rspb.2013.1780.

Chiappe, Luis M. "The First 85 Million Years of Avian Evolution." *Nature*, no. 378 (1995): 349–355.

Chiappe, Luis M., Anusuya Chinsamy, and Peter Dodson. "Mesozoic Avian Bone Microstructure: Physiological Implications." *Paleobiology* 21, no. 4 (1995): 561–574.

Chiappe, Luis M., and Gareth Dyke. "The Beginnings of Birds: Recent Discoveries, Ongoing Arguments and New Directions." In *Major Transitions in Vertebrate Evolution (Life of the Past)*, edited by Jason S. Anderson and Hans-Dieter Sues, 303–336. New York: Columbia University Press, 2007.

Forster, C. A. "The Theropod Ancestry of Birds: New Evidence from the Late Cretaceous of Madagascar." *Science*, no. 279 (1998): 1915–1919.

Foth, Christian, Helmut Tischlinger, and Oliver W. M. Rauhut. "New Specimen of *Archaeopteryx* Provides Insights into the Evolution of Pennaceous Feathers." *Nature*, no. 511 (2014): 79–82.

Louchart, Antoine, and Laurent Viriot. "From Snout to Beak: The Loss of Teeth in Birds." *Trends in Ecology & Evolution* 26, no. 12 (2011): 663–673.

O'Connor, Jingmai K., Luis M. Chiappe, and Alyssa K. Bell. "Pre-modern Birds: Avian Divergences in the Mesozoic." In *Living Dinosaurs: The Evolutionary History of Modern Birds*, edited by Gareth Dyke and Gary Kaiser, 39–114. Hoboken, NJ: John Wiley, 2011.

O'Connor, Jingmai K., Luis M. Chiappe, Cheng-Ming Chuong, David J. Bottjer, and Hailu You. "Homology and Potential Cellular and Molecular Mechanisms for the Development of Unique Feather Morphologies in Early Birds." *Geosciences* 3, no. 2 (2012): 157–177.

Schweitzer, Mary H., Frankie D. Jackson, Luis M. Chiappe, James G. Schmitt, Jorge O. Calvo, and David E. Rubilar. "Late Cretaceous Avian Eggs with Embryos from Argentina." *Journal of Vertebrate Paleontology* 22, no. 1 (2002): 191–195.

You, Hai-lu, Matthew C. Lamanna, Jerald D. Harris, Luis M. Chiappe, Jingmai O'Connor, Shu-an Ji, Jun-chang Lü, Chong-xi Li, Xing Zhang, Kenneth J. Lacovara, Peter Dodson, and Qiang Ji. "A Nearly Modern Amphibious Bird from the Early Cretaceous of Northwestern China." *Science* 312, no. 5780 (2006): 1640–1643.

AVIAN FLIGHT

Henderson, Carroll L., and Steve Adams. *Birds in Flight: The Art and Science of How Birds Fly*. Minneapolis, MN: Voyageur Press, 2008.

Pennycuick, Colin J. *Modelling the Flying Bird*. Burlington, MA: Academic Press, 2008.

Videler, John J. *Avian Flight*. Oxford: Oxford University Press, 2007.

BECOMING AIRBORNE

Burgers, Phillip, and Luis M. Chiappe. "The Wing of *Archaeopteryx* as a Primary Thrust Generator." *Nature*, no. 399 (1999): 60–62.

Dececchi, T. Alexander, Hans C. E. Larsson, and Andrew Allen Farke. "Assessing Arboreal Adaptations of Bird Antecedents: Testing the Ecological Setting of the Origin of the Avian Flight Stroke." *PLoS ONE* 6, no. 8 (2011): DOI:10.1371/journal.pone.0022292.

Dial, K. P. "Wing-Assisted Incline Running and the Evolution of Flight." *Science* 299, no. 5605 (2003): 402–404.

Dial, Kenneth P., Ross J. Randall, and Terry R. Dial. "What Use Is Half a Wing in the Ecology and Evolution of Birds?" *BioScience* 56, no. 5 (2006): 437–445.

Garner, Joseph P., Graham K. Taylor, and Adrian L. R. Thomas. "On the Origins of Birds: The Sequence of Character Acquisition in the Evolution of Avian Flight." *Proceedings: Biological Sciences* 266, no. 1425 (1999): 1259–1266.

Heers, Ashley M., and Kenneth P. Dial. "From Extant to Extinct: Locomotor Ontogeny and the Evolution of Avian Flight." *Trends in Ecology & Evolution* 27, no. 5 (2012): 296–305.

FLIGHT REFINEMENT

Jose Serrano, Francisco, Paul Palmqvist, and José Luis Sanz. "Multivariate Analysis of Neognath Skeletal Measurements: Implications for Body Mass Estimation in Mesozoic Birds." *Zoological Journal of the Linnean Society* 4, no. 173 (2015): 913–928.

Sanz, José L., Luis M. Chiappe, Bernardino P. Pérez-Moreno, Angela D. Buscalioni, José J. Moratalla, Francisco Ortega, and Francisco J. Poyato-Ariza. "An Early Cretaceous Bird from Spain and Its Implications for the Evolution of Avian Flight." *Nature*, no. 382 (1996): 442–445.

THE JEHOL BIOTA

Amiot, Romain, Xu Wang, Zhonghe Zhou, Xiaolin Wang, Eric Buffetaut, Christophe Lécuyerd, Zhongli Ding, Frédéric Fluteau, Tsuyoshi Hibino, Nao Kusuhashi, Jinyou Mo, Varavudh Suteethorn, Yuanqing Wang, Xing Xu, and Fusong Zhang. "Oxygen Isotopes of East Asian Dinosaurs Reveal Exceptionally Cold Early Cretaceous Climates." *Proceedings of the National Academy of Sciences* 108, no. 13 (2010): 5179–5183.

Barrett, Paul M., and Jason M. Hilton. "The Jehol Biota (Lower Cretaceous, China): New Discoveries and Future Prospects." *Integrative Zoology* 1, no. 1 (2006): 15–17.

Chang, Su-Chin, Haichun Zhang, Paul R. Renne, and Yan Fang. "High-precision 40Ar/39Ar Age for the Jehol Biota." *Palaeogeography, Palaeoclimatology, Palaeoecology* 280, nos. 1–2 (2009): 94–104.

Guo, Zhengfu, Jiaqi Liu, and Xiaolin Wang. "Effect of Mesozoic Volcanic Eruptions in the Western Liaoning Province, China on Paleoclimate and Paleoenvironment." *Science in China Series D: Earth Sciences* 46, no. 12 (2003): 1261–1272.

He, Huaiyang, Xin Wang, Zeng Hao Zhou, Fei-Fei Jin, Fu-yun Wang, Lie-kun Yang, Xiaoli Ding, Ariel Boven, and Rongxuan Zhu." The 40Ar/39Ar Dating of the Early Jehol Biota from Fengning, Hebei Province, Northern China." *Geochemistry, Geophysics, Geosystems* 7, no. 4 (2006): DOI:10.1029/2005GC001083.

Jiang, Baoyu, George E. Harlow, Kenneth Woheltz, Zhonghe Zhou, and Jin Meng. "New Evidence Suggests Pyroclastic Flows Are Responsible for the Remarkable Preservation of the Jehol Biota." *Nature Communications*, no. 5 (2013): DOI:10.1038/ncomms4151.

Jiang, Baoyu, and Jingeng Sha. "Preliminary Analysis of the Depositional Environments of the Lower Cretaceous Yixian Formation in the Sihetun Area, Western Liaoning, China." *Cretaceous Research* 28, no. 2 (2007): 183–193.

Ren, Dong. "Flower-Associated Brachycera Flies as Fossil Evidence for Jurassic Angiosperm Origins." *Science* 280, no. 5360 (1998): 85–88.

Zhou, Zhonghe, and Yuan Wang. "Vertebrate Diversity of the Jehol Biota as Compared with Other Lagerstätten." *Science China Earth Sciences* 53, no. 12 (2010): 1894–1907.

Chiappe, Luis M., Shu-An Ji, Qiang Ji, and Mark A. Norell. "Anatomy and Systematics of the Confuciusornithidae (Theropoda: Aves) from the Late Mesozoic of Northeastern China." *Bulletin of the American Museum Novitates*, no. 242 (1999): 836–839.

Chiappe, Luis M., Jesús Marugán-Lobón, Shu'an Ji, and Zhonghe Zhou. "Life History of a Basal Bird: Morphometrics of the Early Cretaceous Confuciusornis." *Biology Letters* 4, no. 6 (2008): 719–723.

Chiappe, Luis M., Ji Shu'an, and Ji Qiang. "Juvenile Birds from the Early Cretaceous of China: Implications for Enantiornithine Ontogeny." *American Museum Novitates*, no. 3594 (2007): 1–46.

Chiappe, Luis M., Bo Zhao, Jingmai K O'Connor, Gao Chunling, Xuri Wang, Michael Habib, Jesús Marugán-Lobón, Qingjin Meng, and Xiaodong Cheng. "A New Specimen of the Early Cretaceous Bird *Hongshanornis longicresta*: Insights into the Aerodynamics and Diet of a Basal Ornithuromorph." *Peer J* (2014): DOI:10.7717/peerj.234.

Chinsamy, Anusuya, Luis M. Chiappe, Jesús Marugán-Lobón, Gao Chunling, and Zhang Fengjiao. "Gender Identification of the Mesozoic Bird *Confuciusornis sanctus*." *Nature Communications* 4 (2013): DOI:10.1038/ncomms2377.

Clarke, Julia A., Zhonghe Zhou, and Fucheng Zhang. "Insight into the Evolution of Avian Flight from a New Clade of Early Cretaceous Ornithurines from China and the Morphology of Yixianornis Grabaui." *Journal of Anatomy* 208, no. 3 (2006): 287–308.

Gao, Chunling, Luis M. Chiappe, Qinjing Meng, Jingmai K. O'Connor, Xuri Wang, Xiaodong Cheng, and Jinyuan Liu. "A New Basal Lineage of Early Cretaceous Birds from China and Its Implications on the Evolution of the Avian Tail." *Palaeontology* 51, no. 4 (2008): 775–791.

Gao, Chunling, Luis M. Chiappe, Fengjiao Zhang, Diana L. Pomeroy, Caizhi Shen, Anusuya Chinsamy, and Maureen O. Walsh. "A Subadult Specimen of the Early Cretaceous Bird *Sapeornis chaoyangensis* and a Taxonomic Reassessment of Sapeornithids." *Journal of Vertebrate Paleontology* 32, no. 5 (2012): 1103–1112.

Li, Dongsheng, Corwin Sullivan, Zhonghe Zhou, and Fucheng Zhang. "Basal Birds from China: A Brief Review." *Chinese Birds* 2, no. 1 (2010): 83–96.

Liu, Di, Luis M. Chiappe, Yuguang Zhang, Alyssa Bell, Qingjin Meng, Quiang Ji, and Xuri Wang. "An Advanced, New Long-legged Bird from the Early Cretaceous of the Jehol Group (Northeastern China): Insights into the Temporal Divergence of Modern Birds." *Zootaxa* 3, no. 3884 (2014): 252–266.

O'Connor, Jingmai K., Ke-Qin Gao, and Luis M. Chiappe. "A New Ornithuromorph (Aves: Ornithothoraces) Bird from the Jehol Group Indicative of Higher-level Diversity." *Journal of Vertebrate Paleontology* 30, no. 2 (2010): 311–321.

O'Connor, J. K., C. Sun, X. Xu, X. Wang, and Z. Zhou. "A New Species of *Jeholornis* with Complete Caudal Integument." *Historical Biology* 1, no. 24 (2012): 29–41.

O'Connor, Jingmai K., Xuri Wang, Luis M. Chiappe, Chunling Gao, Qingjin Meng, Xiaodong Cheng, and Jinyuan Liu. "Phylogenetic Support for a Specialized Clade of Cretaceous Enantiornithine Birds with Information from a New Species." *Journal of Vertebrate Paleontology* 29, no. 1 (2009): 188–204.

O'Connor, Jingmai K., Xiaoli Wang, Corwin Sullivan, Xiaoting Zheng, Pablo Tubaro, Xiaomei Zhang, and Zhonghe Zhou. "Unique Caudal Plumage of *Jeholornis* and Complex Tail Evolution in Early Birds: A New Ornithuromorph (Aves: Ornithothoraces) Bird from the Jehol Group Indicative of Higher-level Diversity." *Proceedings of the National Academy of Sciences* 110, no. 43 (2013): 17404–17408.

O'Connor, Jingmai K., Xioting Zheng, Xiaoli Wang, Yan Wang, and Zhonghe Zhou. "Ovarian Follicles Shed New Light on Dinosaur Reproduction during the Transition towards Birds." *National Science Review* 1, no. 1 (2014): 15–17.

Wang, Xiaoli, Jingmai K. O'Connor, Xiaoting Zheng, Min Wang, Han Hu, and Zhonghe Zhou. "Insights into the Evolution of Rachis Dominated Tail Feathers from a New Basal Enantiornithine (Aves: Ornithothoraces)." *Biological Journal of the Linnean Society* 113, no. 3 (2014): 805–819.

Zhang, Fucheng, and Zhonghe Zhou. "A Primitive Enantiornithine Bird and the Origin of Feathers." *Science* 290, no. 5498 (2000): 1955–1959.

Zhang, Fucheng, Zhonghe Zhou, and Michael J. Benton. "A Primitive Confuciusornithid Bird from China and Its Implications for Early Avian Flight." *Science in China Series D: Earth Sciences* 51, no. 5 (2008): 625–639.

Zheng, Xiaoting, Larry D. Martin, Zhonghe Zhou, David A. Burnham, Fucheng Zhang, and Desui Miao. "From the Cover: Fossil Evidence of Avian Crops from the Early Cretaceous of China." *Proceedings of the National Academy of Sciences* 108, no. 38 (2011): 15904–15907.

Zheng, Xioting, Jingmai K. O'Connor, Fritz Huchzermeyer, Xiaoli Wang, Yan Wang, Xiaomei Zhang, and Zhonghe Zhou. "New Specimens of *Yanornis* Indicate a Piscivorous Diet and Modern Alimentary Canal." *PLOS ONE* 9, no. 4 (2014): DOI:10.1371/journal.pone.0095036.

Zhou, Zhonghe, Julia Clarke, and Fucheng Zhang. "Insight into Diversity, Body Size and Morphological Evolution from the Largest Early Cretaceous Enantiornithine Bird." *Journal of Anatomy*, no. 212 (2008): 565–577.

Zhou, Zhonghe, and Fucheng Zhang. "From the Cover: Discovery of an Ornithurine Bird and Its Implication for Early Cretaceous Avian Radiation." *Proceedings of the National Academy of Sciences* 102, no. 52 (2005): 18998–19002.

Zhou, Zhonghe, and Fucheng Zhang. "A Long-tailed, Seed-eating Bird from the Early Cretaceous of China." *Nature*, no. 418 (2002): 405–409.

Zhou, Zhonghe, and Fucheng Zhang. "Mesozoic Birds of China—a Synoptic Review." *Frontiers of Biology in China* 2, no. 1 (2007): 1–14.

Zhou, Shuang, Zhonghe Zhou, and Jingmai K. O'Connor. "Anatomy of the Basal Ornithurine Bird *Archaeorhynchus spathula* from the Early Cretaceous of Liaoning, China." *Journal of Vertebrate Paleontology* 33, no. 1 (2012): 1–43.

INDEX

evolution: of beaks, 203; of bills, 157, 171, 270; of bony tails, 39, 41, 226–27, 249, 266; of brains, 203; of digestive systems, 157, 188; diversification and, 6–7, 16, 39, 41, 196–97, 264, 271; experimentation and, 213, 221, 226–27; of feathers, 202–3, 214–22, 226, 277; of flight, 39, 41, 95, 108, 199, 214, 227–28, 232, 243–47, 248–53; genetic origin studies and, 238–41, 276; geographical, 6, 193–94; of hands, 144; mass mortality events and, 7, 29, 263, 268; miniaturization and, 73, 108, 157, 201, 209, 251, 253; paleontological gaps in, 16, 36, 105, 196, 227, 255, 264, 276–77; of teeth, 228; of wings, 41, 277. *See also* ecological differentiation; origin of birds

evolutionary transitions: of bony tail, 39, 41, 226, 227, 249, 266; from dinosaurs to birds, 195–96, 198, 225–27; between enantiornithines and ornithuromorphs, 97, 274; Jurassic-Cretaceous, 39

extinctions: at end of Mesozoic, 63, 204, 238, 240, 264; at end of Paleozoic, 195; of feathers, 135, 214, 216, 272; mass, 63, 195, 204, 238, 240, 264; of non-avian dinosaurs, 5, 7–8, 63, 200–202, 204, 240, 277; of primitive birds, 7, 63, 110, 130, 139, 175, 179, 229, 238, 249

eye orbits, 114, 116, 175

fabricated additions, 23

falcons, 139, 242

feathers: aerial competence and, 132, 139, 149, 244–46; aerodynamics and, 41, 153, 208, 209, 214, 217, 234, 266; of *Archaeopteryx*, 224–25, 248; colors of, 130, 220–22, 224; of *Confuciusornis*, 47, 53, 124, 127, 135, 149, 153, 219, 267; of dinosaurs, 202–3, 204–14, 215–22, 245, 255, 258, 260; display and, 37, 47, 69, 124, 127, 132, 149, 207–8, 214, 217, 219–21, 267, 270, 272; diversity of, 132, 213; downy, 204, 207–9, 214–15, 232; of enantiornithines, 63, 127, 135, 149, 153, 216, 231, 270; of *Eoconfuciusornis*, 53; evolution of, 202–3, 214–22, 226, 277; extinct, 135, 214, 216, 272; of hatchlings, 116; of *Hongshanornis*, 99; for insulation, 132, 135, 153, 214–15, 217, 219; leg, 58, 153, 209; maneuverability and, 146, 214, 234, 246, 251–52; ornamental, 43, 47, 53, 69, 127, 135, 149, 212–13, 221, 266–67, 269–72; of ornithuromorphs, 149, 153; of oviraptorosaurs, 207–9; protofeathers, 204–9, 212–14, 216–18, 220; of *Protopteryx*, 69; of *Rahonavis*, 225; roles of, 132, 217, 219, 221; of *Sapeornis*, 55, 58, 269; vs. scales, 215–16; structure of, 208–9, 214–17, 220–21, 244, 270; of theropods, 207, 219; warm-bloodedness and, 217–19; of *Yanornis*, 101, 149; of *Zhouornis*, 91. *See also* flight feathers; tail feathers; wings

feeding behaviors, 185, 228; of enantiornithines, 171, 270–71; ground foraging, 33, 55, 91, 97, 171, 229–30, 237, 240, 267, 275; hunting, 55, 91, 236, 270; of ornithuromorphs, 171; pecking, 236, 275; probing, 77, 87, 171, 232, 236, 240, 270, 275. *See also* diets; digestive systems; foods

feet: of *Archaeorhynchus*, 97, 183; of *Confuciusornis*, 183, 267; of dromaeosaurids, 246; of early vs. modern birds, 23, 179; of enantiornithines, 67, 179, 232–33, 270–71; of *Gansus*, 185; of jeholornithids, 225, 266; lifestyles and, 179, 229; of ornithuromorphs, 272, 274; of *Protopteryx*, 69; of *Rahonavis*, 225–26; of *Sapeornis*, 269; semipalmate, 101; webbed, 179, 185, 236, 240; of *Yanornis*, 101; of *Zhouornis*, 91

ferns, 196, 254, 257

fingers: of *Archaeopteryx*, 248; of *Confuciusornis*, 267; of dinosaurs, 205, 213; of enantiornithines, 230, 250, 270; of jeholornithids, 266; of *Sapeornis*, 269

fish, 196–97, 254, 256, 258–59. *See also under* foods

flamingos, 77, 114

flight, avian, 241–53; arboreal vs. cursorial theory of, 243–45; bony tail and, 33, 43, 248–50; breastbone and, 248; control and, 37, 41, 211, 214, 231, 246, 249, 251–52, 267; efficiency and, 242; evolution of, 39, 41, 95, 108, 199, 214, 227–28, 232, 243–47, 248–53; flap-bounding, 252; forelimbs and, 139, 199, 248; gliding, 139, 211, 213, 242–43, 260, 269; hovering, 243; maneuverability and, 240, 242–43, 246, 249–50; reproductive system and, 108; shoulder bones and, 139, 244–45; size and, 73, 244, 251, 253; speed and, 37, 139, 143, 146, 242–43, 248–52; wing stroke and, 37, 199, 244–45, 247–48, 250–51, 253; wrist bones and, 199, 244, 253. *See also* aerial competence; aerodynamics

flight feathers, 19, 149, 153, 214–15; of *Archaeopteryx*, 223, 248; of early vs. modern birds, 144, 208–9, 216, 231, 251; of enantiornithines, 216, 231, 250–51; of *Jeholornis*, 37; of *Microraptor*, 209–10; of *Rahonavis*, 225; of *Sapeornis*, 55. *See also* feathers

flight muscles: of *Archaeopteryx*, 248; of *Confuciusornis*, 250, 267; of enantiornithines, 67, 69, 270; of maniraptorans, 199; of ornithuromorphs, 234, 252, 274; of *Protopteryx*, 69; of *Sapeornis*, 58

flowering plants, 7, 196–97, 254, 256–58

flycatchers, 149

flying squirrels, 213

foods: fish, 77, 83, 87, 101, 167, 229, 236, 238, 260, 270, 275; fruits, 33, 47, 55, 99, 163, 229–30, 240, 269; grains, 33, 167, 229, 232, 236; insects, 69, 77, 83, 99, 232, 240, 260, 262, 270; invertebrates, 99, 171, 232, 236, 240, 270, 275–76; plants, 262; prey, 55, 91, 236, 270; seeds, 47, 55, 99, 163, 229–30, 240, 262, 267, 269; vertebrates, 236, 240; worms, 232, 236, 262, 270, 275

forelimbs: of *Confuciusornis*, 229, 250, 267; of dinosaurs, 205, 213, 217; of enantiornithines, 250, 270; flight and, 139, 199, 248; of jeholornithids, 37, 266; of maniraptorans, 199; of ornithuromorphs, 234, 252; of *Sapeornis*, 55, 58, 229, 269; of *Zhouornis*, 91